DEVELOPMENT OF LEARNING STRATEGIES WITHIN CHEMICAL EDUCATION

DEVELOPMENT OF LEARNING STRATEGIES WITHIN CHEMICAL EDUCATION

Dr. Terence McIvor

AuthorHouse™ UK
1663 Liberty Drive
Bloomington, IN 47403 USA
www.authorhouse.co.uk
Phone: 0800.197.4150

© 2015 Dr. Terence McIvor. All rights reserved.

No part of this book may be reproduced, stored in a retrieval system, or transmitted by any means without the written permission of the author.

Published by AuthorHouse 02/19/2015

ISBN: 978-1-4969-9498-1 (sc)
ISBN: 978-1-4969-9499-8 (e)

Print information available on the last page.

Any people depicted in stock imagery provided by Thinkstock are models, and such images are being used for illustrative purposes only.
Certain stock imagery © Thinkstock.

This book is printed on acid-free paper.

Because of the dynamic nature of the Internet, any web addresses or links contained in this book may have changed since publication and may no longer be valid. The views expressed in this work are solely those of the author and do not necessarily reflect the views of the publisher, and the publisher hereby disclaims any responsibility for them.

Contents

Abstract vii

Chapter 1   Introduction 1

    Thesis Statement 1
    The Purpose of the Study 1
    The Objective of the Study 1
    Background of the Study 2
    Apparatus and Materials in Chemical Science Teaching 3
    Secondary-school Chemical science in General Education 3
    Related Problems in a Chemical science Program 4
    Need for Experiences with Apparatus and Materials 6
    Learning by Doing 7
    Making Principles and Applications Meaningful 7
    Critical Thinking 7
    Initiative, Resourcefulness, Cooperation 8
    Individual Differences 8
    Variety and Motivation 9
    Types of Experience with Apparatus and Materials 9
    Demonstrations 10
    Laboratory Work 11
    Individual and Group Projects 12
    Visual and Auditory Materials 13
    Use of Inexpensive Apparatus and Materials 13
    Choice of Method 14
    Need for Readily Available Resources 15
    Resource Units 15

Chapter 2 Literature Review 17

    Approaches Aimed at the Development of Reasoning 19
    High-Technology Approaches 20
    Emerging Methodologies: Unstructuring
    The Student Procedures 24
    Main Objective of Laboratory 28

Chapter 3 Methodology 31

    The Introductory College Chemistry Course 32
    The Task Force on General Chemistry 34
    Chemistry in Context 34
    Chemistry Research Theory 35
    Piaget and Chemistry Teaching 36
    Constructivism: A Theory of Knowledge 37
    Application of the Perry Model to General Chemistry 38
    Critical Areas in Chemical Education 40
    Practical laboratory Philosophy 41
    Practical laboratory Methodology 41
    Practical laboratory Curriculum 42
    Analytical Discussion Remarks 44

Chapter 4 Laboratory Experiences in Chemical Science Teaching ... 45

    Purposes of Demonstrations 46
    Demonstrations for Motivation 46
    Demonstrations of Principles and Applications 47
    Demonstrations as Previews 47
    Demonstrations for Teaching for Thinking 48
    Demonstrations for Student Needs 48
    Demonstrations for Skills 49
    Demonstrations for Review 49
    Demonstrations for Evaluation 49
    Criterions for Good Demonstrations 50
    Large-sized Apparatus 51
    Use of Projections 52
    Shifting the Demonstration or the Students 52

        Purposes Not Achieved Automatically 53
        Demonstrations Supplement Other Experiences 53
        Student-performed Demonstrations 54
        Fixed Demonstration Tables 54
        Fixed Demonstration Table Not Essential 56
        Movable Tables for Demonstrations 56
        Nature of the Movable Table 57
        Ceiling Support 57
        General Nature of the Demonstration Equipment 58
        Lighting for Demonstration 58
        Sources of Demonstrations 59
        Laboratory vs. Demonstrations 60
        Individual Laboratory Work 61
        Use of the Laboratory Manual 61
        The Laboratory for Experimentation 62
        Uses of the Individual Laboratory 63
        Sources of Procedures 63
        Laboratory Arrangement 65
        Organization of Laboratory Facilities 65
        Adaptation of Existing Facilities 67

Chapter 5   Improvement in Laboratory Outcome of Teaching 69

        Thinking Not Automatically Achieved 70
        Laboratory Work as Experiences in Thinking 70
        What Is Thinking and Teaching for Thinking? 71
        General Abilities 72
        Special Abilities and Skills 73
        Using Demonstrations for Teaching for Thinking 76
        Use of the Laboratory in Teaching for Thinking 77
        Suggestions for Teaching Plan of Practical Chemistry 78
        Providing for a Variety of Student Problems 79
        Arranging Laboratory Facilities 79

Chapter 6   Usage of Audio Visual Material 81

        The Motion Picture Usefulness 81
        Operation of the Motion-picture Projector 82

Lantern Slides and Projectors 82
Use of the Overhead Projector 84
Constructing an Overhead Projector 84
Materials to Be Projected 86
Demonstration Equipment for Projection 86
Wave Apparatus 86
Electrolytic Cell 86
Polarized Light 87
Kinetic Theory 87
Air-flow Cell 88
Magnetic Field 88
Magnetic Compass 89
Electrical Instruments 89
Sky Charts 89
Use of the Horizontal Lantern-slide Projector 90
The Projection Cell 90
Convection Cell 90
Relative Density 90
Electrolytic Cell 91
Capillary Action 91
Surface Tension 91
Chemical Reactions 91
Special Arrangements 92
Double Reflection 92
Use of Short-focus Lens 92
Built-in Projector 93
Handmade Slides 93
Coating Glass Slides 94
Ground-glass or Etched Slides 95
Protecting Prepared Glass Slides 96
Making Cellophane Slides 97
Filmstrips 98
Opaque Projection 99

Chapter 7 Conclusion 103
References 111

Abstract

The scientific enterprise has had a rich history of producing new knowledge in laboratory settings, and one often pictures the scientist in a laboratory workroom. The formulation of questions, testing of hypotheses, collection of data, and analyzing of data for explanations of causal relationships have all been part of the scientific process. Therefore, education at both precollege and postsecondary levels has almost always contained some laboratory or field work. Although most of the literature for laboratory and field learning appears to come from the natural sciences, its applications are by no means limited to these areas. The principles of learning in laboratory and field settings described here are quite applicable to many other areas of college instruction, such as anthropology, sociology, architecture, engineering, and education. Unfortunately, the means by which most of our students are taught through laboratory instruction in our colleges and universities has not been consistent with the nature of science. The prevailing mode of laboratory instructions has been to have students follow recipe like procedures to verify scientific concepts previously conveyed in the lecture sections of a course.

CHAPTER 1
INTRODUCTION

Thesis Statement

The current status of chemistry practical education and laboratory experience for the first year students is not as encouraging and useful as to produce excellent results. Consequently, in this context, it needs modification to implement a Capstone, addresses the category on Development and Testing of Chemistry Courses or Learning Units and demonstrates an approach to implementing a multi-week capstone project as a part of the laboratory curriculum.

The Purpose of the Study

This study will conduct a thorough research for the innovation in chemistry setting by modifying the content of traditional first-year chemistry experiments and having the students complete a capstone project that addresses multiple content areas. It will also elaborate an assessment of a Materials Development Model, addresses the Development and Testing of Chemistry-Based Instructional Materials category.

The Objective of the Study

This study determines the effectiveness of a materials development model in producing a publishable quality student monograph and instructor's guide to be used in chemical technology education. The materials development model and the resultant instructor's guide could be used in a chemical technology education curriculum.

Background of the Study

The verification method of laboratory learning in higher education is so prevalent that students are usually required to buy laboratory manuals that tell them everything they are to do. There are three major problems with this instructional approach. One is that there is little focus or advance organization for the student. This prevents the student from recognizing relevant versus irrelevant information or procedures at appropriate times during the investigation. It also tends to prevent the building of a conceptual framework to which the student can connect new laboratory experiences. Because there is so much information and procedure, most of what the student does becomes meaningless because the appropriate mental connections are not made. If the student does not know where to go with the instructional plan, he or she is unlikely to arrive them. The second problem is that the procedures in a laboratory "manual" are so laden with jargon and detailed instructions that about all the student can attend to is following the directions. The human brain is capable of manipulating only so much input at any given time, and this threshold is often exceeded while a student is working from a cookbook laboratory exercise.

There is frequently neither time during the period nor any provision for thinking about the scientific questions being asked or about the procedural options, for analyzing the data, or for developing specific scientific concepts and chemical science process skills. The student perceives the objective of the laboratory experience as finishing the procedure rather than attaining specific learning objectives. Even though the task required is usually completed in the required time, it is questionable as to what concepts and process skills are acquired.

The third problem is that this instruction tends to be so "numbing" that it not only fails to teach students process skills and specific course concepts but it also becomes tediously boring. It is no secret that most students in non major laboratory classes do not enjoy working in the lab.

Such laboratory experiences contribute to students' dislike of science. Fortunately, the above does not describe all laboratory programs in our colleges and universities. There are some appropriate examples of rich chemical science learning experiences for our students.

Apparatus and Materials in Chemical Science Teaching

Chemical science has played an outstanding role in our life in recent years and is now changing our entire existence in such important aspects as health, transportation, communication, and power. The social, economic, and political implications are both national and international in character. We are truly living in an air age, a power age, and an atomic age, where boys and girls are literally forced to be concerned with some principles of chemical science and their ever important applications in everyday living. This applies not only to the minor fraction of high-school graduates who go to college, but to all youth living, working, learning, and enjoying hobbies, in the home, on the farm, in the factory, in the office, or on vacation. All are vitally affected by science. All need to be familiar with applications and implications of principles of chemical science to be able to live effectively in a technological world and to be intelligent concerning the complicated problems with which they must deal as citizens in our democratic society.

Secondary-school Chemical science in General Education

Secondary-school chemical science is, therefore, concerned primarily with the general education of all young people, including the approximate four fifths who do not go to college as well as those who do.

The modern secondary school is gradually attempting to formulate its program to meet the needs of the youth who attend it. While preparation for college is recognized as one of the needs of a relatively small group, the nature and method of school experience, as well as its evaluation, are evolving in many well-planned schools in such ways that all may benefit. A recent statement of the National Education Association summarizes this point of view:

Schools should be dedicated to the proposition that every youth in these United States--regardless of sex, economic status, geographic location, or race--should experience a broad and balanced education which will (1) equip him to enter an occupation suited to his abilities and offering reasonable opportunity for personal growth and social usefulness; (2) prepare him to assume the full responsibilities of American citizenship; (3) give him a fair chance to exercise his right to the pursuit of happiness; (4) stimulate intellectual curiosity, engender satisfaction in intellectual achievement, and cultivate the ability to think rationally; and (5) help him to develop an appreciation of the ethical values which should under gird all life in a democratic society. Another group has expressed this point of view by the statement:

The purpose of general education is to meet the needs of individuals in the basic aspects of living in such a way as to promote the fullest possible realization of personal potentialities and the most effective participation in a democratic society. A similar statement has recently been phrased that says that the purpose of the schools is to provide the best possible conditions for the student's steady and harmonious growth both toward his own individuality and toward his more responsible membership in a democratic society.

Related Problems in a Chemical science Program

Any program for chemical science in the secondary school planned for achieving the purposes of general education is concerned with four interrelated problems: the goals or objectives to be met, the content or subject matter to be included, the experiences to be provided or the methods to be used, and the evaluation procedures to be utilized in determining the extent to which progress has been made toward achieving the objectives. These are very closely interrelated and are not separate categories. The content, or subject matter, as well as the methods used, should be suggested by and consistent with the objectives. The evaluation should be concerned with the objectives sought and the procedures used. The chemical science teacher is vitally concerned with all these problems. In the actual day-to-day work with students, however, certain problems recur: What experiences can and should be provided? What activities are most appropriate to achieve the purposes which have been agreed upon? These are problems of method, of procedures, of planning and organizing student activities. What are the students to do, and what does the teacher do to encourage and guide those experiences adequately?

The present study is concerned with only the latter phase of these interrelated problems; particularly with that part of method and student activity related to the use of apparatus and materials. Student experiences with chemical science materials should be present invariably in any chemical science course. What the objectives should be, what particular courses and kinds of experience should be provided, how these should be organized, what subject matter topics should be included, how the program should be evaluated--all are problems outside the scope of the present treatment.

There are implications for objectives, content, method, and even evaluation in the specific suggestions which are made relating to the use of apparatus and materials. It seems impossible to think of worthwhile, major objectives, whether for a chemical science course for a special group preparing for college or for the unselected majority in a general education program that would not be furthered by appropriate experiences with chemical science apparatus and materials.

It is, therefore, the purpose here to present suggested ways of obtaining and using the large variety of equipment needed by the modern school, with its provision for individual needs and interests. Decisions concerning the use of demonstrations, laboratory work, student projects, and field trips should be made by the teachers and the students in each situation. Such decisions should be governed by the promise of an experience or a procedure in promoting the objectives of the course; the objectives are continuously refined in the light of the students' needs and previous experiences. Such other procedures as discussions, problem-solving, directed study and committee work become valuable as they provide experience, and as they help to interpret experience. They should be used whenever they are the most appropriate experiences for the purposes desired.

Need for Experiences with Apparatus and Materials

Because the modern secondary school is concerned not only with providing a background of functional subject matter but also with the development of individuality and responsible participation in our democratic society, experiences and activities for students become increasingly important. The questions with which the chemical science teacher is concerned are: What are students going to do? What activities will they engage in? What experiences should they have? While it may seem to some that firsthand contact with the apparatus and materials of chemical science is obviously essential, there are many present-day schools where the practice belies this belief. In many classes the use of textbooks and recitation procedures is stressed, while the use of demonstrations, laboratory work, and student projects is minimized, if not lacking entirely. Students read and talk about scientific devices and procedures rather than actually work with them.

This is particularly true in general chemical science classes where many teachers and administrators make no apologies for having limited facilities and equipment for direct student experience in the classroom. Even many physics and chemistry classes, in which a laboratory period has become almost traditional, have to a large extent eliminated demonstrations and student projects and instead have provided a stereotyped set of "experiment" which everyone follows from directions in a manual to the extent allowed by inadequate supplies and equipment. There usually develops also a pressure for completing a certain list of experiments, so that there is little or no time for less formalized student experience.

If the desirability of using apparatus and materials of chemical science in secondary-school courses needs supporting arguments or reasons, these may be grouped under the following headings: (1) We learn by doing. (2) Scientific principles and applications can be made meaningful. (3) There are possibilities for teaching for critical thinking. (4) Initiative, resourcefulness, and cooperation may be furthered. (5) Provision may be made for individual differences. (6) Variety and motivation may be provided.

Learning by Doing

Educators have long recognized the advantage of firsthand, direct experience. This is exemplified particularly in chemical science where many of the concepts, as well as the techniques and devices, are new to the beginner in the field. These new terms, principles, and materials are made meaningful by actual use. The laboratory method which has been the very essence of chemical science for many years has been deemed so important that its use has been carried into many other areas of study.

Psychologists have determined, on the basis of findings over a period of years that learning by doing has many advantages over mere reading about principles, concepts, and applications. Student experience with actual materials and phenomena is desirable and necessary in order to understand important facts and principles. Laboratory experiments, individual student projects, and field trips may be made firsthand experience. Demonstrations also may be direct experience, even though in some instances they are performed by the teacher.

Making Principles and Applications Meaningful

Many principles of chemical science are quite abstract to high-school students who are meeting them for the first time. Students try to learn important laws and generalizations by memorizing them. Well-planned and executed demonstrations and firsthand experiences in the laboratory can make important principles and applications more meaningful than oral or written explanation. Perhaps it should be pointed out that demonstrations, laboratory work, projects, and field trips in themselves do not automatically make principles and applications meaningful to the student. Carefully used, they give meaning to what would be otherwise vague abstractions.

Critical Thinking

The use of apparatus and materials in teaching for thinking is vital in chemical sciences pedagogy. However, it should be pointed out here that demonstrations and laboratory work provide excellent opportunities for student experience in reflective thinking. Problems relating to the "how" and "why" and "what would happen if" type of question are possible and pertinent in connection with demonstrations and laboratory experiments. Thinking is involved in the planning, in the interpretation of the result, and in the generalizations possible on the basis of the outcomes.

Initiative, Resourcefulness, Cooperation

The modern school places emphasis upon developing the whole personality of the child, resulting in an important concern for the provision of situations where students, on the one hand, have opportunity for exhibiting initiative and resourcefulness and, on the other, for working together with other students. Laboratory experiments can provide for all these if carefully planned and carried out. This is the place where students participate with the teacher in planning the laboratory activities, the situation calls for initiative and resourcefulness on their part.

The usual laboratory situation with limited equipment necessitating that two or more work together on laboratory experiments engenders cooperation. The use of student projects and special reports is another opportunity for initiative and resourcefulness. Field trips afford opportunity for cooperation, especially where students participate in the planning, arranging of details, and in setting up the particular learning experiences. In handling any apparatus whether with experiments, demonstrations, or special projects, initiative must be shown in thinking through the particular materials needed, in setting up the apparatus, in improvising and substituting for parts which are not available, and in adjusting and repairing the apparatus. This is related to the problem of critical thinking; working with apparatus necessitates thinking in solving the problem for which the apparatus is being used.

Individual Differences

Another emphasis in present-day school programs leads to the attention which should be given to the wide variety of needs and interests in high-school youth. Not all students learn equally well from books.

Some are very much interested in the handling of apparatus. Interests which students have in radio, electricity, chemistry, machinery, and the like may well be discovered, explored, and encouraged in connection with activities involved in demonstrations, laboratory work, and projects.

Variety and Motivation

One of the advantages of chemical science teaching is the range of activities which can be utilized to provide a wide variety of teaching procedures and motivations. There should be little excuse for monotony in the chemical science classroom. The use of apparatus and materials in such situations as dramatic demonstrations, silent demonstrations, experiences in critical thinking, visual material, auditory presentation, and special exhibits suggests something of the range and interesting variety which the chemical science teacher has as resources. Ingenuity and planning prevent the chemical science class from being dull.

Types of Experience with Apparatus and Materials

As emphasized at another point, there is no infallible way to provide experiences suitable for all students with their differing needs and interests. This is as true with respect to the use of apparatus and materials as with any other kind of activity. What is appropriate at one time, for a given group with respect to one kind of objective, is inappropriate at another time, with other students, or for a different goal. It is fortunate, therefore, that there is a considerable array of possibilities for the use of apparatus and materials. For purposes of discussion these may be grouped into the following categories: (1) demonstration; (2) laboratory work; (3) individual and group projects; (4) field trips; (5) visual and auditory materials.

The first four of these are major student learning experiences or teaching procedures. The fifth may be a procedure in itself, or may supplement other procedures. The visual or auditory device becomes the procedure in such instances as the use of slides or filmstrips to illustrate principles, facts, or applications. In other cases a handmade slide may be used with a demonstration, a motion picture with a discussion, a recording with a laboratory experiment, or a filmstrip with a student report.

The various procedures are closely interrelated in actual practice; in fact, their interrelatedness is one criterion of a good situation. However, they are sufficiently distinct to consider separately. Since there has been some lack of agreement in terminology and associated meaning, it is desirable to indicate briefly what is meant by each of these five types of experiences with apparatus and materials.

Demonstrations

Demonstrations involve the use of apparatus and materials, or both, and are usually performed by a single individual before a group. Often these are carried out by the teacher but may be performed by one of the students or by an assistant if one is available. Demonstrations may be performed to illustrate principles, to teach for critical thinking, to show how a particular piece of apparatus operates, to teach skills, to help clarify certain concepts, or to illustrate important applications. Demonstrations are often qualitative rather than quantitative, and the outcomes or results may be known in advance. For example, one may demonstrate the principle that for a given resistance the greater the applied voltage, the greater is the amperage through the resistance. This can easily be demonstrated to a group with an appropriate voltmeter, ammeter, and resistance visible to all by attaching leads with spring clips first to a single cell of a storage battery, then to two cells, and finally to an entire 6-v battery. Another example of a typical demonstration is the illustrating of various effects of air pressure by a series of simple applications, such as water held in an inverted tumbler with a card, rise of liquids in a soda straw, the need for two holes in emptying a can of condensed milk, the principle of the mercurial barometer, and the crushing of a can by atmospheric pressure.

Examples of demonstrations of other principles include those of osmosis with a carrot and glass tube, the preparation and properties of oxygen from mercuric oxide, the electrolysis of water, or the operation of a thermostat. Demonstrations may profitably include quantitative work, which provides opportunities for the use of quantitative instruments, skill in use and reading, and for critical thinking in the interpretation of data.

What ordinarily would be an individual laboratory experiment may, of course, be used as a demonstration before the class. This may be either qualitative or quantitative. The demonstration of Boyle's law is an example of the latter. This may be done by the teacher or by a student. A J-tube may be used, and the readings placed on the blackboard as mercury is added. Projects, developed in or out of class, may also be demonstrated. Examples of such projects are the operation of a mock-up of an auto ignition system or a working model of a Cottrell precipitator. The major distinguishing characteristics of a demonstration are that they are shown before a group, usually by the teacher or a student, and they involve the use of apparatus, or materials.

Laboratory Work

High-school laboratory work usually involves the carrying out of experiments by individuals or by small groups. Usually there is a problem for which an answer is being sought. While in many cases a laboratory manual or workbook is followed, such need not be the case. Frequently all students in the class work on the same experiments at the same time, although it is often very desirable that more than one kind of experiment be undertaken by different students or groups. In an individual laboratory experiment each student manipulates the apparatus and carries out the steps necessary to find answers to definite questions.

Laboratory experiments in the high school may be either quantitative or qualitative. Examples of the latter include the preparation of hydrogen and a study of its properties, and the connection of bells or lights in series and in parallel, with the advantages and limitations of each. Examples of quantitative experiments include the determining of the linear coefficients of expansion of different metal rods, or finding the percentage of acetic acid in a sample of vinegar.

The distinguishing characteristics of laboratory experiments are that they are performed with apparatus and materials by an individual or a small group to find either a qualitative or a quantitative answer to a problem.

Individual and Group Projects

The term project has been used in a rather wide variety of ways. For the present purpose it may be considered simply as a problem upon which a student or a small group works. This problem may be of a laboratory type, in which case the project is essentially the same as a laboratory experiment. The project may be a problem which would not involve the use of what is ordinary termed apparatus. The making of a chart, the preparation of materials for a bulletin board, or the preparation of a series of handmade slides are examples of this kind of project. The term special report may be applied to an individual project concerned with such problems as the biography of a noted scientist, the historical development of a scientific device, or the discussion of the practical applications of an important principle.

The term project is here used as it relates to carrying out a problem concerned with the construction, repair, improvising, or use of some apparatus, device, or process. Project work may be done either in or out of class. Project work may be given or reported upon in class or outside of class, as to a chemical science club, a parent teacher meeting, or school assembly. In some cases it may not be presented or reported at all, except perhaps to the teacher. A student who rewinds a car generator to make a motor may simply use it and not report upon it. A group project resulting in a chemical science mural or a public address system for the classroom may likewise need little if any reporting. The performing of a demonstration or an experiment may also constitute a student project.

The obvious difference between an individual project and a group project is simply that the latter is carried out by a group instead of by an individual. Sometimes this means that different students carry out certain parts of a project which can be put together into a whole at the completion; in other instances all the groups work together on the whole project. Examples of the former would be constructing a series of slides to illustrate the development of the airplane or the parts of an automobile engine. Examples of the latter type would be the construction of a radio set or a model sulfuric acid plant.

Visual and Auditory Materials

Visual materials include such devices as motion pictures, filmstrips, glass and cellophane slides, opaque projectors, maps and charts, pictures, models and mock-ups, and the blackboard. Other teaching materials may be included in the category of visual; the term is used here to include only the special visual materials which have not been included under other categories. Demonstrations and laboratory experiment, for example, are in one sense visual materials but are not included in the present use of the term. Auditory materials include the use of the radio, public-address system, phonograph, and recorder. Such a device as a sound motion picture is, of course, a combination of auditory and visual equipment.

Use of Inexpensive Apparatus and Materials

In many cases, the most important outcome in connection with apparatus and materials is obtained from the experience in their use, even when the results are quantitative; thus, the equipment is not necessarily elaborate or expensive. While some pieces of apparatus such as balances, voltmeters, and motors should be purchased, much of the experimental work may be carried out satisfactorily with inexpensive equipment improvised from materials from the home, ten-cent store, or salvage store. Students can bring materials and devices from the home: tin cans, wire from discarded devices, motors, pumps, bottles, parts of autos or tractors, boxes, radios, and the like. There are psychological advantages in the use of familiar materials, and in the fact that they have been supplied by the students themselves. Such inexpensive equipment is often more effective than commercially made equipment. The latter, with its fine finish, elaborate arrangement, technical appearance, and its working parts obscured, is often less realistic and functional than that made from familiar material.

Good learning experiences are a desirable outcome of the planning, improvising, and repairing of homemade equipment. Here, too, are opportunities for providing for the varied individual differences and interests. It should be pointed out that some good commercially made apparatus is also desirable and necessary; this is particularly true where corresponding equipment cannot be made easily. A supply of usable commercial equipment should be built up in a school over a period of years. A wide range of desirable student experience can be obtained only when there is an adequate supply of both commercial and homemade equipment.

Choice of Method

Since there is such a wide range of differences in students, their backgrounds, goals, interests, parental desires, and the like, it is clear that there can be no one best way of teaching for every group. This is as true in providing experiences in the use of apparatus and materials as with other procedures. Many teachers look hopefully to some "expert" or to educational research for the best procedure for teaching. Since teachers also differ from each other, procedures which are effective for one are not necessarily satisfactory for another. If all the factors are quite similar; if the purposes are the same, the students of similar interests and abilities, with similar previous experiences; and if the methods of different teachers are identical, perhaps best procedures can be indicated. This is the only situation in which it is advisable to suggest that certain procedures are superior to others, e.g., that demonstrations are better than laboratory experiences, slides are better than motion pictures.

For many, if not most, goals of chemical science teaching, direct experience through the use of apparatus and materials is desirable. The way in which this is to be accomplished should be determined in each situation, whether by means of demonstrations, individual laboratory experiments, student projects, or field trips. Probably some experiences of each type should be provided, particularly over an extended period of time.

Need for Readily Available Resources

To provide worth-while student experiences requires careful planning. Since there is no blueprint for procedure, plans for experiences for each group of students have to be made anew each time. The same courses cannot be provided for different students year after year using the same assignment, laboratory experiment, motion pictures, guide sheets, and tests, and really provide experiences appropriate to their varying needs. There is, therefore, need for a wealth of materials, procedures, devices, and suggested activities with which to plan.

Resource Units

One way of making available such a wealth of materials and suggestions is by means of resource units. While these have undergone some changes in theory and practice in the past few years in the hands of different authors and teachers, 6 the fundamental idea involved is simple and functional. As the term suggests, a resource unit is simply the accumulation in one place (notebook, filing-cabinet drawer, shelf) of all the resources which could be used for teacher and student experiences in connection with one of the major units into which the course or area has been organized. A resource unit on communications, for example, may contain suggested laboratory experiments, demonstrations, student projects, field trips, workbooks, films, slides, charts, recordings, problem sets, pamphlets, clippings, questions to be raised, guide sheets, bibliographies, problems involving reflective thinking, tests, and other materials which could be used in teaching this unit. These would constitute or represent the resources available, from which choice could be made in providing the experiences for a particular group of students.

Not all the specific possibilities assembled in the resource unit would be utilized at any one time, but they would be at hand for possible use to meet the wide range of interests and needs of different groups. Resource units should be continuously evolving. New materials should be added even at times when the unit is not being taught, and obsolete or unworkable material discarded. New references, free and low-cost pamphlets, articles from current magazines, films and recordings, are new resources which keep the resource unit in a continual state of growth, and the teacher on the alert to keep it growing and up to date. Much of the content of this volume, though not arranged in the form of resource units, can be utilized as resource materials.

Chapter 2
LITERATURE REVIEW

Systematic observations of Higher Secondary school science-laboratory classes in a large Midwestern university during the late 1990s revealed that students were, for the most part, performing cookbook-like laboratory experiments and were not learning the process skills of science. Instead, students spent a significant amount of laboratory time listening to the instructor transmit information. These observations are probably typical of most university laboratory courses in the sciences. Also the use of investigative laboratory teaching strategies at the college level has lagged behind those used in the secondary and elementary schools (Kyle et al., 2000).

During the past decade, some interesting experimental studies developed and tested alternative approaches to the traditional laboratory investigation. Most of the innovative approaches are modifications of the inquiry model and employ discovery or inductive approaches to learning. Utilization of science-process skills--such as hypothesis formation, identification and manipulation of experimental variables, and inferring from data--are characteristic of these newer approaches.

In one such study, students in a physical-science class for elementary education majors employed concrete, manipulative laboratory experiences to learn about measurement, pressure, Archimedes' principle, machines, and electricity. This class showed greater achievement on some of the concepts, a greater comprehension of all areas taught, and better attitudes toward science than a similar class taught without the manipulative laboratory experience (Splickler, 2001). Kem and Carpenter (2001) found that a field-oriented, onsite, instructional approach to geology produced significantly more interest and enjoyment than a traditional laboratory approach. Ilhe laboratory method associated with a learning-center approach to college geology was found to have a significantly greater effect on short-term learning than did a more conventional laboratory approach (Tofte, 1999).

A number of studies on laboratory approaches have been done in introductory biology. Lawson and Snitgen (1999) found that a series of laboratory investigations to teach formal reasoning were found to improve significantly the ability of students to use formal operational thought. Similarly, cognitive development of college, non-biology-major students was found to be promoted by a laboratory program that emphasized investigation and that took into account the limitations of students' cognitive ability (Journet, Young, Stanley, & Scheibe, 1998). Walkosz and Yeany (2001) found that training in integrated scienceprocess-skill development improved the performance of college biology students in the use of integrated science-process skills. A successful inquiry strategy developed originally for the high-school level by the Biological Sciences Curriculum Study (BSCS) was adapted for use in a university general-biology laboratory program and tested experimentally for an entire semester against a well-established commercial program that was highly directive

(Leonard, 1999). The BSCS approach made systematic use of science processes, of development of concepts via questioning, and of requirements of the student to make procedural decisions.

The commercial approach required primarily the following of instructions exactly as stated and the answering of a few very specific questions. The scores of students using the BSCS orientation were similar to those using the commercial approach on a pretest of selected biology laboratory concepts, but their scores were significantly higher on a posttest. This study was later replicated with students at two small, private colleges. Again, the group using the BSCS orientation scored significantly higher on a test of biological laboratory concepts at the end of the semester (Hall & McCurdy, 2001). Finally, an extended-discretion laboratory approach in which students we're required to determine their own investigatory procedures and strategies was tested experimentally against a more directive laboratory approach in university general biology. Even though the former approach placed much greater demands on student creativity and decision making, student scores between the two groups on learning measures given at the end of the semester did not differ significantly. One conclusion is that students can learn at least as much when given fewer procedural directions on laboratory investigations (Leonard, 2000a).

Studies on the use of investigative or inquiry approaches in college science laboratory courses suggest that inquiry laboratory strategies (1) are more student involved and more inductive than traditional approaches, (2) contain less direction and give the student more responsibility for determining procedural options, (3) require students to make more extensive use of science-process skills, (4) produce significantly greater educational gains than traditional approaches, and (5) appear to work equally well for college students of all ability levels, not just the very academically talented.

Students appear to prefer inquiry style instruction as well. A survey of six hundred students in introductory, nonmajor science courses showed that the students had a clear preference for investigative laboratory activities to the standard, structured activities (Davis & Black, 2000). Inquiry laboratory programs have been found to be workable. For example, research supports recommendations for maintaining the spirit of process-oriented laboratory investigation, for removing the cookbook from commercial laboratory investigations, and for helping laboratory instructors become better teachers (Leonard, Journet, & Ecklund, 1999; Leonard, 2001).

Approaches Aimed at the Development of Reasoning

Some very interesting work on the development of reasoning and problem solving abilities has been inspired by the theories of developmental psychologists such as Jean Piaget. Lawson (1998) has argued that many major concepts typically taught in introductory science courses require formal operational thought in order to be understood. He found that approximately one-half of students in college introductory biology did not fully understand formal operational thinking and that these students were unable to develop an appreciable understanding of abstract concepts. Fields (2000) found that typical, traditional (cookbook) laboratory exercises did little to foster the understanding of abstract concepts. There is evidence that students can be taught to improve their use of formal operational thought. Inquiry-based, hands-on approaches in general biology laboratory courses wherein a series of laboratory investigations to teach formal reasoning formed the major mode of instruction were found to improve significantly the ability of students to use formal operational thought (Lawson & Snitgen, 1999; Journet et al., 1998).

Lawrenz (2000) found that grouping students in a physical-science laboratory class for elementary education majors by similar cognitive ability was more effective--as far as their achievement was concerned--than other alternative forms of grouping. Intensive instruction on cue attendance was found to significantly increase the problem-solving abilities of science-methods students, especially those skills related to developing and testing hypotheses (Wright, 1999). Very diverse populations of college students have been taught strategies for using formal reasoning in their science courses (Morgan et al., 1998). It is believed that, in order to teach concrete or transitional thinkers formal operational concepts, they need first to work with a principle in concrete terms, especially with manipulative activities provided in a laboratory environment.

The students can then be supplied with symbolic representations of the principles so that the principles can be progressively internalized through the gradual reduction of the need for perceptual and motor supports: this provides students with an opportunity to gradually abstract the principle from its concrete exemplars (Lawson & Wollman, 1997; Lawson, 2000).

High-Technology Approaches

The microcomputer introduces several applications of computer-based technology to laboratory instruction in college science courses. Almost as soon as the microcomputer was used for science instruction, faculty, especially those in the physical sciences, began the development of computer-based applications. Currently there are two major uses of computer-based technology in college laboratory courses: (1) for direct instruction of laboratory concepts by simulation, using traditional computer-assisted instruction (CAI) or a more advanced version of CAI with an interactive videodisc system (IVI), and (2) for data analysis or input of data with laboratory instrumentation interfaced to a microcomputer. The following discussion is divided into these two areas.

There are numerous studies using CAI in college science instruction, but there are only a few reports of use in conjunction with laboratory instruction specifically.

In one study, students in an introductory chemistry laboratory course who used computer-simulated experiments for four different laboratory investigations (kinetics, absorbance spectroscopy, emission spectroscopy, and equilibrium) did as well or significantly better than students using traditional laboratory techniques for the same topic. The CAI group also spent significantly less time learning the material (Calvin & Lasgowski, 1998). Curtis (1999) used a software system designed to teach students how to adapt simple response functions to experimental data. Modern data-analysis techniques were found to help students with levels of achievement. Miller (1999) found no differences in achievement or attitudes owing to student use of CAI materials in a community-college biology laboratory course.

Microcomputers interfaced to laser videodisc players provide a combination of the advantages of microcomputer and traditional television or videodisc images. The result of interfacing these two technologies permits a high level of interaction between the computer and student, as well as high-resolution, lifelike video images of natural phenomena (Leonard, 2000b). In a comparison of interactive videodisc versus the traditional laboratory technique for teaching physical principles of standing waves and strings, no difference was found on pretest/posttest gains between the two groups of students, except that students in the two groups used different strategies to separate and control for variables based on the physical nature of the instructional materials available (Stevens, 2000). Waugh (1998) randomly assigned two groups of chemistry students studying equilibrium to either a traditional laboratory activity or simulation with an interactive videodisc system. The latter group scored significantly better on both laboratory quizzes and on their laboratory reports.

Similarly, a large group of non-biology-major students were assigned to either traditional laboratory exercises or simulations on an interactive videodisc system to learn about cellular respiration and about biogeography. Results showed no statistically significant differences between the groups on laboratory quizzes, laboratory reports, or laboratory final exam. Opinion data on a questionnaire indicated that students felt the videodisc instruction gave them more experimental and procedural options and more efficient use of instructional time than did conventional laboratory instruction. Students indicated that interactive videodisc was equivalent for general interest, understanding of basic principles, help on examinations, and attitudes toward science. The conclusion was that interactive videodisc can, in some cases, provide comparable instruction to the wet laboratory (Leonard, 2000b, 2002).

One of the most exciting developments in laboratory instruction is the interfacing of laboratory measurement devices to a microcomputer. Nicklin (2000) found that many physiological experiments could be improved and made more accurate by interfacing common physiological instruments to a microcomputer. He also found that the microcomputer could act as a "lab partner" for students working individually on an experiment and that interfacing was not expensive. Old kymograph transducers interfaced with microcomputer-based workstations for undergraduate physiology laboratories were found to be very functional and successful (Rhodes, 1999). Morgan, Markell, and Feller (1998) have given a complete description of interfacing muscle-physiology measuring devices to a microcomputer. One of these is a pistol grip transducer that is used to study contraction of the human trigger-finger muscles. An excellent and illustrative guide for inexpensively constructing interfaces for twelve common laboratory instruments--such as a thermistor, motion time, pH meter, and humidity meter--has been prepared by Vernier (1998).

A simple and inexpensive interfacing kit, called Science Tool Kit, is available from Carolina Biological Supply and other science-supply companies. The basic module for the Apple II sells for $70 and contains experiments in biology, chemistry, and physics. Additional modules for speed and motion, earthquakes, and human physiology are available for $40, each with additional experiments. A variety of other commercial interfacing kits are available as well. For example, IBM is developing a Personal Science Laboratory (PSL) that can be used in college science laboratories. There are educational benefits of using instruments interfaced to a microcomputer in the laboratory. These benefits include reducing cost, improving effectiveness, saving student time (thus preventing boredom), learning to use state-of-the-art scientific instrumentation, simplifying data analysis, making experimental results more meaningful by allowing students to perceive relationships between independent and dependent variables as the experiment is completed, allowing students to more effectively comprehend abstract concepts, and providing opportunities for developing problem-solving skills (Leonard, 1998, 2003).

Ideas for classroom interfacing come from scientific research. Among the ideas being developed in research that may have interesting applications for the classroom are trackers for eye, head, and hand gestures; tracers of eye direction and focus tracking; and voice recognition and synthesis (Foley, 1998). IBM has an interactive system capable of recognizing 20,000 words (98% of the typical English-speaking vocabulary). The development of much more powerful microcomputers, CD-ROM, and image capturing by microcomputers will soon be available for classroom use. Future possibilities for laboratory interfacing are almost unlimited.

The recent development and research on applications of computer technology for laboratory instruction in college-science courses suggests that applications of computer technology in the laboratory classroom is here to stay and that science faculty will continue to develop new applications for instruction. The temptation to tinker with this new technology is almost irresistible. The demonstrated educational benefits of computer applications for student learning also appear to be equivalent to or better without computer-based laboratory instruction.

Inferences from the Literature on Laboratory Learning

Inquiry or investigative approaches in college-laboratory-science courses appear to be quite productive. Newer and more innovative approaches over the past ten years have been more student-involved, more inductive, and have required more extensive use of science-process skills. These new approaches have generally produced significantly greater educational gains than the more traditional approaches. There is some evidence that students can be taught to improve their use of formal operational thought through the use of concrete, manipulative, laboratory experiences. If given appropriate laboratory experiences, some students designated as concrete thinkers can develop an understanding of concepts that are considered to require formal thought. The more direct the student involvement in all aspects of the laboratory activity, the more the student appears to learn. Computer-based applications in the laboratory are of great interest to college science faculty and have, in some cases, proved to be as productive or more productive than the conventional laboratory exercise.

Emerging Methodologies: Unstructuring The Student Procedures

Give the student a simple task or goal to accomplish. This means more than stating the title of the investigation or telling the student the purpose of the investigation. An example for the popular experiment concerning light intensity and the photosynthesis of elodea is, "Find the effects of different light intensities on the rate of photosynthesis in elodea." The task can be both printed and reiterated verbally. The student should see this as the objective of the laboratory period rather than as "getting through the manual." Moreover, it is helpful to have the students restate the task in their own words and discuss it briefly in small groups before proceeding further.

- Give the student only essential procedures. This can be accomplished by reading through an existing cookbook investigation and striking all procedures or explanations that you think are nonessential or that you feel the students can figure out themselves. As a rough guide, about half of the existing verbiage can be eliminated from most commercially available laboratory exercises. Have these essential procedures retyped so that they can be provided to the student. Title them "Suggestions" and arrange them sequentially.

Make each "suggestion" relatively brief and simple so that each can be fully internalized by the student. The result will be a series of statements that do not detail exact procedures but that provide clues about what tasks must be done. The remainder is up to the students. Encourage students to figure out some of the procedural details themselves. Make sure that some of these procedures call for the collection and analysis of some data. If hypotheses are appropriate, have students construct them as well.

- Have students work in small, cooperative groups. This will provide several sources of ideas on how to work out some of the specific logistics. Because the students have to think through some of these procedures for themselves and because they must interact with other students, they become more actively involved in the investigation both physically and mentally. Provide the student with a list of discretionary resources. These should include:

- A list of the materials available to carry out the task. Students should be able to visualize all the materials available and their possible association with the task.

- A list of special technical procedures. These detailed prescriptions--such as how to stain a slide or how to measure acceleration--need not be invented by the student and should be available when the student determines that they are needed. This provides meaningful relations between the goal of the investigation and the prescribed procedures. More importantly, the focus remains on the main goal or question of the investigation rather than on the technical procedures. Do not tell the student when technical procedures are to be used, but identify clearly what each is designed to do. When a student selects a procedure, even if it is an inappropriate one, at least he or she has given some thought as to why that one was selected.

Ask some meaningful questions at the end of the investigation. Most cookbook laboratory investigations are like shaggy-dog stories in that they lead the student through a procedure but fail to develop a relation between the investigation and the concept(s) to be learned because they lack the necessary follow up questions. Three kinds of questions are helpful here. First, ask the students to summarize the data or to examine the data for patterns or regularities. Then ask questions that will allow the students to infer from the data possible answers concerning the task or goal. Questions asking for support or rejection of hypotheses are appropriate here. Finally, ask extrapolation questions or ones that suggest generalizations and implications. This gradual and hierarchical process of thinking through the data facilitates conceptualization. It is based on the assumption that certain levels of learning require corresponding prerequisite learning.

Resist telling the student how to carry out the investigation. Give the students help in other than technical procedures only when you feel that they are becoming so frustrated that it is unlikely they will be able to determine a reasonable plan on their own. Coach and coax them without giving it all away. There is nothing wrong with students failing to always make the "right" decisions. Permit the students to make mistakes, since these often result in valuable learning.

There appears to be emerging a theory of science learning that is closely connected to laboratory instruction. The theory will be identified, for lack of a better label, as active teaming. It is becoming clearer in educational research that learners who are actively engaged in the instructional process are the most successful learners.

Most experienced instructors have for some time had an intuitive feeling that if a student is directly involved (physically, emotionally, and mentally) with the concepts or skills to be learned, this student will have a deeper understanding--and will retain that understanding longer than if the learning experience had been passive. This is one reason why the lecture, although superficially expedient, is not a very effective teaching strategy, particularly for students who have not developed the ability to easily learn abstract concepts. That is why some college-level laboratory programs have attempted to provide students with a rich array of experience-based instructional experiences, including laboratory investigations, projects, visuals, interactive reading, and inquiry discussions.

Laboratory inquiry is central to active learning because it tends to be student centered--and, therefore, more directly experienced. The learner's mind must be actively engaged for inquiry to occur. All students, not just the academically talented, can be successful inquirers. A very pragmatic view of inquiry is, "Try to let the students figure out the concept, rather than tell it to them." Yes, less "material" can be "covered" with inquiry instruction than with lecture-based instruction. The instructor must make a decision about how to trade off quality with quantity.

A great deal of what is traditionally taught in introductory college science courses is quite abstract. Some examples in biology are mechanisms such as evolutionary theory, respiration, photosynthesis, protein synthesis, and recombinant DNA. In chemistry, orbital theory, bonding, quantum mechanics, and even dimensional analysis are complex and abstract. In physics, almost everything except mechanics is abstract and complex. Certainly most theories in science are abstract concepts. Current estimates are that approximately half of the freshman college population nationally are not fully capable of formal operational thought Should this concern us? If it concerns you, consider the following hypotheses which have the support of modern cognitive research.

1. Most learners will benefit from concrete learning experiences prior to their receiving indirect instruction about abstract science concepts. A lecture of an abstract nature and reading should follow, rather than precede, an engaging activity related to the concepts to be learned. It appears that all but the most motivated students learn very little meaningful science from lecture and textual reading alone. This hypothesis suggests that a productive learning sequence is probably
(1) orientation, (2) hands-on investigation, (3) discussion--perhaps a little lecture--then (4) reading and working on problems. The support for this hypothesis probably begins with the work of Piaget but is confirmed more recently in studies by Renner, Lawson, Abraham, and others.

2. Learners must reconstruct new knowledge of our culture as if it were entirely new to them. Most knowledge, if it is to be applied, cannot simply be imparted (poured into a student's

head). Learners must interact with and reconstruct the concepts for themselves. Not all of this needs to be with concrete objects, but there should be meaningful interaction between the learner and what the learner is to acquire. The concept behind this hypothesis has been given the popular label "constructivism."

3. Learners attempt to connect new conceptual development to their existing cognitive framework. This suggests that providing the student with a conceptual framework and advance organizers that fit onto the framework will allow the student to fit what is being learned into what is already known. If the connection is successful, that knowledge is more easily retrievable, is more lasting, and is able to be more meaningfully applied in other contexts. The concept behind this hypothesis is often referred to as "connectivism."

Main Objective of Laboratory

One of the main objectives of laboratory or field education is to provide the student with the opportunity to learn how to think. It is my opinion that knowledge of the processes of learning is a more important skill for the student than knowledge of the content of science. This means thinking about connections between the major concepts of a course, the goals in a given laboratory investigation, and the use of science-process skills. Knowledge of science processes will allow the student to better learn the traditional content and to make these important connections.

Less prescriptive approaches to learning science have been found to be quite successful in college-level laboratory courses for non majors (Case, 1998; Lawson & Snitgen, 1999; Fields, 2000; Leonard, 1999; Journet, et al., 1998; Nisbett, Fong, Lehman, & Cheng, 1998; Lawson, 1999; Leonard, 1999; Abraham, 2000; Lawson, Rissing, & Faeth, 2003; Leonard, 2000a, 2000c; Hall & McCurdy, 2001). Programs which foster the development of critical thinking and other higher-order thinking skills are approaches that are predominant in twelve exemplary college science programs selected by the National Science Teachers Association (Crow, 2000).

Further, more "open-ended" or indirect approaches for laboratory instruction are being strongly recommended by major commission reports such as Project 2061: Science for All Americans (American Association for the Advancement of Science, 2000), Fulfilling the Promise: Biology Education in the Nation's Schools (National Research Council, 2003), and The Liberal Art of Science (American Association for the Advancement of Science, 2003). The following passages from the latter source argue on this behalf.

Thus, use of the confirmatory approach in the laboratory and in the field does not contribute to the development of strong conceptual links between the natural world and the scientific theories developed to explain and predict it. Nor does this practice leave students with an accurate view of the practice of science. Rather, it contributes to the notion that the purpose of experimentation is the verification of hypotheses rather than their refutation.

Maximum benefit can be derived from laboratory and field experiences by having students work in groups and share their ideas, perceptions, and conceptions. Group design and interpretation of laboratory work am also effective strategies for exposing the changing misconceptions. In addition, students should prepare written reports describing the rationale for the experimental design, the data, and their interpretations.

Since the recommendations by national commissions for changes in science education at the college level are grounded in learning research, we must take these recommendations seriously if our students are to appreciate and be literate in science. College and university science instructors, especially those involved in course or curriculum development, are encouraged to implement in their institutions.

These faculty members are further encouraged to report in the literature their experiences so that the entire science-education community may benefit from the creative efforts of others. Some specific recommendations on how to conduct research on instruction in a college or university setting are given in a series of three articles in the Research and Teaching column of the Journal of College Science Teaching (Leonard, 2002; Abraham, 2004;). The authors of these articles address strategies for asking the right research questions, design of instructional experiments, and analysis of data. Expanding the knowledge base in college instruction will serve to give recognition to instructional pedagogy as a legitimate field of scientific inquiry in itself.

CHAPTER 3

METHODOLOGY

According to the July 2001 statement of the American Chemical Society, "Education Policies for National Survival," the study of chemistry is essential if students are to understand the natural world and is the key to success in a variety of careers both within and outside the chemical sciences. No group understands these facts better than undergraduate chemistry professors (in both two-year and four-year institutions) who face an exciting challenge each term. This challenge is personified in a diverse student population that has a broad range of preparative skills and career objectives. Many students enter their undergraduate chemistry courses with no experience in chemistry at all or, at most, a one-year high school course in chemistry.

Many enter so-called non major courses in chemistry with the intention of fulfilling a requirement, whether for a liberal-arts degree; for engineering, nursing, or physical-therapy degrees; or for the fulfillment of requirements for premedical, predental, and biological-science programs. A very small minority enters these courses as declared majors in chemistry or biochemistry. In every case, the prior chemical experience of these students is very uneven, ranging from minimal to no chemistry to several years of guided research.

Although recent literature in chemical education tends to give major attention to introductory chemistry courses designed for the students identified above, chemistry education also takes place in the subsequent years of the undergraduate curriculum, from courses in organic chemistry and physical chemistry to advanced courses that serve both undergraduate and graduate students. For these courses, the Committee on Professional Training of the American Chemical Society (ACS) has formulated guidelines to assure minimal competency in these areas.

The Introductory College Chemistry Course

The attention given to introductory college chemistry is well deserved and will be the focus of the remainder of this part. Indeed, Project Kaleidoscope, in its monograph "What Works: Building Natural Science Communities," states unequivocally that the transformation of introductory courses must be the National Science Foundation's highest priority over the next five years since a significant body of research confirms that the first year of college is the critical drop-off point in numbers of students in science and mathematics courses and that students acquire and confirm lifelong beliefs and attitudes about science and mathematics in their introductory courses. Until recently, very few alternatives to the "traditional" general chemistry course could be found in the undergraduate chemistry curriculum either in major or non major courses. Brock Spencer, in his 2001 Chemical Manufacturers Association Catalyst Award address, "What Works in Chemistry Education," aptly described this traditional course

> an unrelated set of problems at the end of this week to be worked by choosing the right formula to apply, a three-hour lab to be endured in which the purpose is to come as close as possible to the 'right' answer, and an occasional multiple choice exam.

> Students view the material as a set of isolated exercises to be solved rather than as part of an exciting conceptual structure, as preparation for the next course rather than as preparation to understand the world, and as an impersonal, competitive and isolating experience. (Spencer, 2001)

The non major course was often taught as simply a less rigorous version of the major course, but with the same emphasis, as Sheila Tobias has reported in "They're Not Dumb, They're Different" (2003) regarding dry, factual, predigested rule-ordered material. And all this in lecture halls containing two- to five hundred students and in laboratories (when they exist) supervised by graduate students with minimal instruction in pedagogical skills. That's the bad news. The good news is that something is being done about it. Spencer (2001) has cited the many different models that are currently being tried in many kinds of institutions. These models include:

- New introductory courses that pose and solve real experimental problems in an investigative approach to chemistry using modern instruments;

- Courses that provide direct laboratory experience with phenomena first, followed by development of concepts based on those investigations;

Courses that coordinate content with other science courses, so that students see connections among several disciplines;

- courses that feature varied methodologies such as peer learning, concept development rather than topic coverage, classroom attention to current scientific literature, and societal and personal issues.

The initiatives described by Spencer come at a time that the ACS, through the agency of some of its branches, is attempting to support curricular change in introductory chemistry courses. Recognizing that undergraduate courses must support the educational needs and career aspirations of future citizens as well as future scientists, the ACS (American Chemical Society, 2001) has recommended that

- Introductory chemistry courses be redesigned in order to rekindle student interest and address the specific needs of unprepared students;

- Current efforts to develop a scientifically literate citizenry be expanded;

- The reward structures of colleges and universities include incentives for quality instruction and provide opportunities for faculty growth and development;

- Strategies and incentives be developed to encourage the entry of underrepresented populations into chemistry;

- Major laboratory curriculum development address the needs of undergraduate institutions to upgrade instructional equipment, provide hands-on chemistry experiences for all undergraduates, and establish more undergraduate research opportunities for chemistry majors.

Some Specific Initiatives in Chemistry Education

Two specific ACS (2001) initiatives, one addressing the major course and one addressing the non major course, am described below.

The Task Force on General Chemistry

Several years ago, recognizing the critical need to address the specific issue of general chemistry courses for the career-related disciplines of the majors and sciences, Stanley Kirschner, then chair of the Division of Chemical Education, ACS, established a Task Force on General Chemistry. Members of the Task Force believed that general chemistry needed to be revitalized and were concerned with both content and process of what was taught and how it was taught. The Task Force was composed of three subcommittees to carry out its charge from three different points of view: a modular/core curricular approach, a subject-area approach, and a laboratory based approach. Although the Task Force agreed that there was not one "best" approach to teaching general chemistry, it also felt that general chemistry needed to be unburdened from its present abundance of detail and incoherence. The Task Force's focus was to develop a curriculum suited to all introductory chemistry students in two-year and four-year colleges.

Chemistry in Context

The ACS is also sponsoring the development of a new college chemistry textbook entitled Chemistry in Context: Applying Chemistry to Society, edited by A. Truman Schwartz (2004), intended primarily for students who do not anticipate majors or careers in chemistry or other sciences. The primary goal of Chemistry in Context is to motivate students to learn chemistry so that they can and will act as responsible citizens in our increasingly technical age. The text has been structured in such a way that students will discover the theoretical and practical significance of chemistry and will also become aware of what a very human endeavor chemistry is. In this textbook a broad range of group and individual activities has as its object the empowerment of students so that they can locate information, and develop analytical skills, critical judgment, and the ability to assess risks and benefits.

In our advanced scientific and technological society, we must pay special attention to the science and technology education of the non-specialist. Our scientific enterprise and, indeed, our well-being as a society will be doomed unless we quickly develop a literate citizenry--one that can distinguish between astronomy and astrology; that can deal successfully with the complex issues related to animals' rights; that can benefit from advances in the nutritional sciences; that can deal responsibly with pollution and pollution control; and that can appreciate the benefits of chemicals, their potential hazards, their safe handling, and their disposal. The very democratic principles upon which our society was founded and continues to function are now seriously threatened and will be jeopardized unless we achieve a state of literacy in science, mathematics, and technology. Literacy in those fields is a measure of our values as a society: what we are about, what we believe in, how we treat each other, and how we treat our plane.

Chemistry Research Theory

Over the past twenty years, research in chemistry teaching has revealed that a vast majority of chemistry students at all levels, including the graduate level, learn chemistry concepts by rote and solve chemistry problems by using algorithmic methods. Although many students perform satisfactorily on exams, it has been found that interviews with students can reveal gross misconceptions regarding chemical phenomena (Bodner, 1999). The insights into student learning outlined below can help instructors rethink the teaching process so that they can teach for meaning and not simple rote playback of chemical concepts.

Piaget and Chemistry Teaching

Herron (1997) breakthrough paper applies Piaget's theories on how we acquire knowledge to the teaching of chemistry. Piaget distinguished among four stages of intellectual development: the sensory-motor, preoperational, concrete operational and formal-operational stages. The concrete-operational student structures and organizes activity in reference to concrete things and events in the immediate present. Such a student does not think in terms of possibilities and is not able to understand abstract concepts that depart from concrete reality. The formal-operational student, on the other hand, thinks--or at least is beginning to think--in terms of what might happen and envisions all the changes that are possible. Formal-operational students can reason without the aid of visual props.

Although the instructional approach that we take--and virtually every concept that we teach--in chemistry requires learners to be at the formal operational level (normally reached by age fifteen, according to Piaget) if they are to comprehend the concepts that are presented, a widely publicized study done at the University of Oklahoma indicated that only 50 percent of the college freshmen who were tested functioned completely at the concrete operational level and that only 25 percent of the sample could be considered fully formal in their thought processes. Such statistics should have tremendous influence on how we teach chemistry.

Herron has suggested that we confront the problem of delivery of chemical concepts to concrete-operational students in one of two ways: either skirt the problem or overcome it. Skirting the problem involves making formal concepts accessible to concrete-operational students by emphasizing concrete concepts and testing for their mastery. Overcoming the problem involves taking steps to enable concrete conceptualizers to develop into formal thinkers at some later time. One suggestion that Herron has offered is that we can help students acquire surrogate concepts that can substitute for the real thing by providing extensive experience with concrete props that model the abstract concept. The hope is that the transition from the surrogate to the real will become increasingly easy as the student matures. This transition can be encouraged if students are forced to think about what they are doing, are engaged in the intellectual debate of ideas, are required to weigh evidence, and are helped to make sense of a series of observed facts. Although provision of these educational experiences is often frustrating and time consuming and requires a great deal of interaction among students and between student and teacher, instructors can make considerable progress in chemistry teaching if they take the time to provide some of these experiences.

Constructivism: A Theory of Knowledge

According to Bodner (1999), the constructivist model of learning can be succinctly summarized by the statement: Knowledge is constructed in the mind of the learner. This idea is a logical outgrowth of Piaget's model of intellectual development, since his model was built upon the assumption that knowledge is constructed as learners strive to organize their experiences in the framework of preexisting mental structures. Table 16.1, summarized from Bodner's paper, contrasts the traditional view of learning with the constructivist view. The constructivist model can be summarized as follows:

Table 16.1

Summary of Characteristics of the Traditional View and the Constructivist View of Learning

Traditional View	**Constructivist View**
Reality is a static body of knowledge	Construction is a process in which knowledge is both built and continually tested
Mind is a black box	Environment is a black box
Stimulus-response is accurately judged	Mental process is accurately judged
Mental process is guessed at	Relationship between mental structures and real world is guessed at
Learners mirror and reflect what they hear and read	Learners construct own knowledge by looking for meaning and order
Learning is more passive	Learning is more active

Traditional View	Constructivist View
There is a search for a match with reality	There is a search for a fit with reality (lock and key analogy)
Copies or replicas of reality are in learners' minds	Many keys with different shapes can open a given lock

Source: Bodner, 1999.

- It can be very helpful in explaining and overcoming student misconceptions, which are so resistant to instruction that the only way to replace a misconception is to help students construct a new concept that more appropriately explains the experience.

- It has some important implications for instruction. For the instructor it requires a subtle shift in perspective from someone who "teaches" to someone who tries to facilitate learning: a shift from teaching by imposition to teaching by negotiation.

- It facilitates a two-directional flow of information between student and teacher--which, in turn, requires students to explain their answers, to reflect on their learning process, and to be responsible for the language they use.

- It is a paradigm of the basic scientific-research process.

Application of the Perry Model to General Chemistry

The realization that students are active learners with various stages of intellectual, emotional, and ethical maturity can and should have a profound effect on the methods that instructors use for teaching and the environments that they create for learning. A cognitive development model that seems useful to chemistry educators is that described by William Perry (2000). David Finster (2000, 2001) has shown how the various stages, or positions, of intellectual and ethical Table 16.2 Relationship of Perry Categories to Student Responses to an Introductory Lecture on Chemical Bonding

Perry Category	Student Response
Dualism: The student sees the world in terms of opposites: good-bad, right-wrong, we-they. Truth is absolute; un-certainty is only temporary.	Student enjoyed lecture on chemical bond-ing because professor conveyed air of authority. Student was confused because the advantages and disadvantages of each theory were summarized, but the teacher never said which theory was "right."
Multiplism: Diversity and uncertainty are legitimate; all opinions are equal, in-cluding those of authorities.	Student enjoyed hearing about different approaches to bonding; dilemma is try-ing to guess which approach the teacher thinks is right.
Relativism: Student recognizes that knowl-edge is contextual and relative.	Student used to think that scientists always had a single right answer, but now finds that each theory can be used effectively in a given situation.

Source: Finster, 2000.

development in college students within the Perry model can be related to how students learn chemistry. Perry's scheme of development can be grouped into four categories--dualism, multiplicity, relativism, and commitment to relativism--each of which represents a unique way of thinking or a particular cognitive filter through which students understand their world.

Regarding the first three categories as applied to chemical bonding, Finster has given an example in which he asks us to imagine that a teacher has just finished a traditional lecture on an introduction to Valence Bond Theory and Molecular Orbital Theory, summarizing the advantages and disadvantages of each as applied to the explanation of the properties of the homo nuclear diatomic molecules of Period Two in the Periodic Table. The Perry schema offers a possible framework for understanding the various student responses, as listed below in Table 16.2.

Although Finster recognized that students can be found anywhere along the line of progress within the Perry scheme, the realistic fact is that most freshmen function as dualists and that general chemistry classes are largely filled with dualistic thinkers who expect a dualistic approach.

However, it is appropriate to promote growth along the scheme using the strategy of developmental instruction. Finster has suggested, in a matrix of challenge and support issues for general chemistry classes, ways in which dualists, multiplists, and relativists can be both supported and challenged. For example, dualists are supported by a highly structured course that includes lectures providing clearly defined terms, a detailed syllabus, a clear set of expectations, homework assignments that parallel the text material, and so forth. Multiplists can be supported and dualists can be challenged by organizing the course with some flexibility concerning content and sequencing, by providing some directions about how to generate problem solving strategies, by structuring group work and analysis of laboratory results by groups, and by having students control or design some aspects of the learning experience.

Relativists can be supported and multiplists can be challenged by providing for a more independent learning environment, by encouraging students to develop their own definition of problems and to work out their own solutions, by letting students select their own laboratory problems and modify the design, by providing an historical/societal context for the course content, and by testing across the whole range of Bloom's taxonomy. Other challenge/support issues are diversity of the learning experience, methods of experiential learning, and personalism. While it is obvious that the Perry schema can be an exciting challenge to the traditional way in which chemical educators have "delivered chemistry," instructors must proceed with some caution since each will find himself or herself in a unique learning context with a unique student body.

Critical Areas in Chemical Education

Within the context set forth above, a survey is conducted of approximately a dozen nationally recognized chemical educators. They produced the following "laundry list" of critical areas that they feel must be addressed by the chemistry education community. Their remarks have been grouped under the five headings of philosophy, methodology, curriculum, laboratory, and assessment--although there is a great deal of overlap among these areas.

Practical laboratory Philosophy

- Learning chemistry is a highly personal endeavor that requires the learner to pass judgment on the significance and degree of interest in what is learned.

- Chemistry is the central science that is connected to all other scientific disciplines.

- Learning is a challenge, and learning chemistry is a big challenge--but worth it.

- Conveying a sense of grandeur: chemistry is one of the supreme accomplishments of the human mind.

- Abstract principles can be related to everyday happenings; abstractions are not understood unless they can be applied.

- Modern chemical theory evolved through development of models that have continually been perfected through experimental observation.

Practical laboratory Methodology

- Teaching chemistry is best approached as a process emphasizing critical-thinking skills and problem-solving skills rather than as the accumulation of memorized information, facts, theories, and algorithms.

- Personal attention to students is very important.

- Meeting students at their level is essential to communication.

- Strategies to cope with verbal and mathematical illiteracy must be devised in cooperation with other campus departments.

- Varied approaches to problem solving help students with varied backgrounds: drill and practice; cultivation of higher order thinking skills; dimensional analysis. A discovery, or guided inquiry, format is often a successful approach.

- Variations on lecture, such as cooperative learning, demonstrations, experimentation, small group discussions, and so forth are essential in optimizing the learning process.

- Emphasis on vocabulary building is very important. Learning chemistry is like learning a new language; time must be taken to learn the language of chemistry in association with direct chemistry experiences so that it can serve as a tool for critical thinking. Avoidance of lecture entirely is an approach advocated by many chemistry educators. A student's attention span in lecture is about ten minutes; passive learning amounts to virtually no learning. As an old Chinese proverb says "Tell me, and I will forget; show me, and I may remember, involve me, and I will understand."

Practical laboratory Curriculum

The curriculum should be rational and reasonable. It should not be designed to "cover" all aspects of chemistry superficially but should be more depth oriented in its approach.

- Many recognized chemistry educators constantly search for creative alternative ways to introduce students to chemistry.

- It is important to include the social and historical framework in which chemistry is an evolving discipline rather than a static body of knowledge, so that chemistry can be perceived as a human endeavor evolving within a social and political context with all the strengths and fallibilities attendant on human beings.

- Spiraling--that is, returning to topics periodically at more and more sophisticated levels--is preferable to broad, superficial coverage of topics.

- Integration of new technology into teaching is essential. Chemistry is "done" using technology, and many areas of chemistry can be better taught using technology.

- Inclusion of the environment as a wellspring from which many chemical examples can be drawn helps to relate the subject matter to the world in which we live.

Laboratory.

- A balance of small-scale and macroscale approaches in the laboratory can give the student a broad range of experience with a variety of techniques.

Emphasis on open-ended procedures and investigative techniques is the hallmark of the constructivist approach in the laboratory.

- It is important to have a balanced approach to safety in the handling of chemicals and apparatus without prescribing so many caveats that the laboratory process is impeded or halted altogether.

- The laboratory is the place where students have the greatest opportunity to really understand what chemists actually do and how they think.

- Problems of budget, safety, and time should be carefully thought out and addressed; they need not impede laboratory instruction if the laboratory is well planned.

Assessment.

- There is need for thoughtful, creative design for assessment instruments.

- Balance in assessing learning is important. Assessment can be overdone by spending excessive time on it; it can be underdone by assessing only very low level skills and knowledge.

- Instructors should include meaningful assessment of laboratory work if they truly believe that it is an essential part of learning and teaching chemistry.

Analytical Discussion Remarks

Whether chemistry educators are involved in teaching introductory courses or more advanced courses, the only thing of which they can be certain is that everything is in a state of flux and is subject to change. Groups of chemistry educators are constantly examining the philosophical underpinnings, assumptions, prerequisites, requirements, and curricula at every level of undergraduate chemistry courses. Such challenges and changes to the status quo are reflected in the contents of the internationally recognized standardized chemistry tests produced by the Examinations Institute of the Division of Chemical Education, ACS. For example, the ACS General Chemistry Examination was once described as presenting chemistry as a collection of facts and equations along dualistic lines: completely objective; dealing with facts, principles, and equations in a multiple-choice format; excluding issues of value, history, and process--and as such a reflection of what chemical educators value (Finster, 2000, 2001).

However, things are changing. The ACS Examinations Institute is now exploring ways to "break the bubble" (of the multiple-choice answer sheet) by designing machine-scored examinations-with more than one correct answer--that challenge students to think in more multiplistic and relativistic terms. A symposium that addresses these new thoughts in testing and evaluation took place at the Twelfth Biennial Conference on Chemical Education in August, 2002. The paper titles alone indicate the direction that assessment is taking: "Evaluating Problem-Solving Proficiency through Performance Assessments," "Hands-On and Minds-On Testing," "Alternative Testing Formats: The ChemCom Experience," "Super Test: A Flexible, Computer-Based Assessment Project."

It is clear that many thoughtful and concerned chemistry educators are working to optimize chemistry classroom/laboratory instruction nationwide. (Moore, 2000). In interviews with many of these educators, it has also become clear that they also realize that it is imperative to treat chemistry instruction in more than one human dimension. The holistic approach to education no longer allows either student or teacher the security of conventional pedagogy, nor does it allow teacher or learner to become manipulators of knowledge while leaving the inner self unexamined. Scientific principles cannot be divorced from the fundamental ethical principles that guide the decisions that necessarily evolve from scientific research.

Chapter 4

LABORATORY EXPERIENCES IN CHEMICAL SCIENCE TEACHING

While only the teacher, or at times one or two students, gives a demonstration, all the students before whom it is given should be having a worth-while experience relative to one or more important goals. In order that these experiences may be really functional, some attention should be given to planning all the activities related to it. Students can derive value from demonstrations only to the extent that these become worthwhile experiences for them, and they are not likely to have worthwhile experiences unless the demonstrations are planned with this in mind. Even then, there are unforeseen difficulties, but these are far fewer than when the demonstration is haphazardly performed rather than carefully planned. Worth-while experiences usually do not occur by chance.

The purpose for which a demonstration is to be given, therefore, is a first and most important consideration. Upon the purpose depends not only the appropriateness of the demonstration and hence whether or not it should be given, but also how it should be presented and what the students do with respect to it.

Purposes of Demonstrations

Like most other effective techniques or procedures in science teaching --motion pictures, laboratory, directed study, discussions, for example-demonstrations may be used for a variety of purposes. They are very versatile. They provide the possibility of a range of student activities. Among these purposes for which demonstrations have proved functional are: (1) motivation; (2) explanation of a principle or application; (3) preview of a unit of work; (4) provision for various phases of teaching for thinking; (5) provision for particular student needs or interest; (6) exemplification of a skill or technique; (7) review; (8) evaluation.

Demonstrations for Motivation

Teachers know that people have a natural curiosity and interest in actual objects and in seeing things in motion. The downtown store window demonstration always attracts passers-by. In the classroom a sample of iron ore or a flowing siphon is an immediate center of interest. The classroom demonstration thus is one excellent way to make science interesting for students. In spite of recent popularizing through the press and screen, and contact with the many practical applications of science in the modern home, even many simple principles and facts of science have much of the magic element for students. While they live in a world of marvels, the scientific explanations are well hidden. The automobile, familiar as it is to American youth, is a good case in point. The driver pushes the horn button, the clutch pedal, the accelerator, the brake pedal, has a spark plug or fan belt replaced, and replenishes the gasoline, oil, and water, but in many instances has not seen or does not recognize the generator, distributor, differential, camshaft, thermostat, carburetor, or starter switch, much less know their function and how they operate.

Students are used to commercial devices that compel their attention and interests, along with attractive provisions which have been made by other agencies in this respect. The motion picture, radio, magazines in color, parks, playgrounds, athletic contest, ready transportation, comic books, and even the store windows and billboards, all are in competition with the school for the students' interest. How drab and uninteresting for students exposed to such lures are the classes which use only the textbook! The least that the science teacher can do is to make the science class and the room interesting and attractive. The demonstration is one good way to motivate.

Demonstrations of Principles and Applications

Probably the most usual use of demonstrations is for illustrating and explaining scientific principles and their applications. For most students, seeing the real thing is much more helpful than reading about it or looking at a picture of it. The latter may be necessary initial activities, with the demonstration used to clarify and exemplify. In some instances, the demonstration may well be used as the introductory experience, followed by such other activities as discussion, reading, films, and laboratory work. While the demonstration may not automatically provide an understanding, it furnishes a real experience upon which the teacher may build, along with other well-chosen procedures and activities.

Demonstrations as Previews

The frequent plunging into a unit of work in the text leaves many students with a sense of bewilderment when they might understand or appreciate the details if they were given some perspective on the unit as a whole. The use of a series of demonstrations at the beginning of a unit of work, even before any reading has been assigned, helps students get an overview of what the unit is about, with some idea of the relationships, important ideas, and applications. Such previews also make the unit more interesting because the students come to realize that they must have some understanding and skills relative to the principles in order to be able to work on the more interesting applications coming later.

Dr. Terence McIvor

Demonstrations for Teaching for Thinking

Since one of the important goals of the modern secondary school is to teach students to do critical, reflective thinking or to use scientific methods, the experiences in science classes should make a major contribution toward this objective. The demonstration is one effective way of providing experiences in thinking. While activities toward this end are not automatic and some little sensitivity and planning on the part of the teacher are necessary, most demonstrations are replete with numerous possibilities for such experiences. The definition of terms, the applications of principles, the identification of assumptions, the collection and interpretation of data, the use and significance of controls, the testing of hypotheses, the drawing of conclusions, the making of accurate observations--one or more of these is inherent in most demonstrations.

Some specific descriptions and suggestions of how demonstrations may be used in teaching for thinking. It may be pointed out here that if the purpose of the demonstration is to include experiences in thinking, the teacher should have this as an objective and plan accordingly for the presentation of the demonstration and the participation of student. Most of the important phases of scientific thinking, as of any other concept or skill or habit, need continual emphasis if they are to be meaningful and usable. Demonstrations offer one means of getting this emphasis with new and interesting problems and situations.

Demonstrations for Student Needs

It is well recognized that all students do not learn equally well from the same experiences. There is always need for a variety of experiences. Demonstrations offer one way to provide variety in the usual routine of class work. Some students need the experience of appearing before a group; assisting the teacher in the demonstration and explaining certain operations provides such experience. Some need to learn to be more accurate in their statements and use of words; an explanation to the class of the meaning of a demonstration and the use of appropriate terminology help to meet this need. Needed skills such as accurate observation, organization of data, and writing an understandable report may be met through the use of good demonstrations.

Demonstrations for Skills

While skills such as weighing, pouring reagents, or operating special equipment are best learned by direct experience, initial as well as later instruction can be given profitably by demonstration. This can be given to a whole group, rather than to an individual. Cautions, points to watch for, evidences of good techniques, and related theory often can be stressed in a more satisfactory way as a group demonstration.

Demonstrations for Review

An effective method of review is by means of a series of demonstrations. These may be a repetition of those which have been given previously, with discussion or other student activity. Demonstrations which have not previously been used may be given to illustrate principles and applications which have been studied. The so-called "silent" demonstration may be used; the teacher demonstrates without comment and the students are asked to give the explanation of the phenomena observed, either orally or in writing.

Demonstrations for Evaluation

Demonstrations can often be utilized to give evidence of ability to utilize certain skills, make observations, interpret data, apply principles, or set up apparatus. Such a test offers variety from the usual pencil-and paper type and, if properly set up, furnishes more valid evidence of these kinds of achievements. Such testing may be used to supplement the traditional type.

Criterions for Good Demonstrations

Reflection upon the purposes which demonstrations may serve leads to the conclusion that there is no single procedure to be followed. Any suggestions or directions which indicate that in giving demonstrations one should always do (or not do) a particular thing, completely disregard the relating of procedure to purpose. However, it is possible to indicate certain criterions for good demonstrations which do apply to most situations. These criterions are desirable but are not patterns or blueprints of procedure. They are guides for arranging and carrying out demonstrations so that they may be most effective. While these criterions are applicable in general, there may be some situations in which one or more is not pertinent. In such a case the particular criterion would be modified or disregarded entirely. The criterions for good demonstrations include:

1. The demonstration should be tried out previously.
2. The purpose of the demonstration should be clear.
3. The demonstration should be visible to all student.
4. The apparatus used should be as simple as possible.
5. The demonstration should be utilized as fully as possible in the light of its purposes.

1. The Demonstration Should Be Tried Out Previously

No demonstration should be given before a class if it has not been tried out previously, preferably a short time before the class period. This applies to the simplest kind of demonstration as well as to elaborate ones. It is true, as well, for the experienced teacher who has used the demonstration before. All the needed apparatus and materials should be secured and prepared in advance so that the continuity is not broken by the search for them. Simple as the diagram in the textbook or manual may appear, there is often an obscure part or particular skill which needs some previous attention. It is quite embarrassing to face a class with a demonstration which does not work.

2. The Purpose of the Demonstration Should Be Clear

Students should know what the demonstration is about, what it illustrates or brings out. Often the haste with which the demonstration is performed, the newness or complication of the apparatus or the lack of familiarity with the principle shrouds the main purpose and point, and renders the demonstration quite ineffective.

3. The Demonstration Should Be Visible to All Students

It may seem superfluous to suggest that students should be able to see the demonstration, and yet this is one of the most common faults of this procedure. Teachers should not assume that because they can see the apparatus clearly all others in the room can do so. A number of factors may contribute to poor visibility, and they suggest the need for attention to the following points. If possible the demonstration table should be lighted so that the apparatus stands out clearly. Shades should be adjusted so that students can see from all parts of the room. If necessary, the students should be allowed to change their seats or stand where they can see.

Large-sized Apparatus

Use should be made of devices which are large enough to be seen from the rear and sides of the room. Apparatus companies are giving considerable attention to this problem, with the result that there is available a quantity of large-sized apparatus which is particularly effective with large classes. This includes such things as glass tubing, beakers, bottles, and test tubes, as well as demonstration motor, electric meters, balances, and gyroscopes.

Use of Projections

Many demonstrations which are difficult to show directly to a large class may be projected on the screen by means of a projection lantern or a spotlight. This is particularly true in the case of numerous electrical and optical demonstrations, such as magnetic fields, electroscopes, shadows, reflection, refraction, and the like. A projection cell may be used very effectively to project numerous demonstrations in chemistry and electrochemistry.

Shifting the Demonstration or the Students

Often the demonstration can be given in a place other than the teacher's desk, so that students can see it better. With simple demonstrations, the teacher may be in front of the desk, or even in the midst of the students. Often he can pass among the students so that all may see the results of the demonstration, such as the precipitate in the test tube, the expanded ball, or that the water does not run out of the bottle. Sometimes it is desirable to give the demonstration first on one side of the room and then on the other. Often it is feasible for students to group themselves around the table while the demonstration is in progress.

4. The Apparatus Used Should Be as Simple as Possible

In general the use of the simplest and most familiar apparatus and devices makes for the most effective demonstrations. The purpose of the experiment may be obscured when complicated equipment (often commercial) is used. Many of the most usable demonstration devices can be made or improvised in simple form.

5. The Demonstration Should Be Utilized as Fully as Possible in the Light of Its Purposes

Teachers often hurry through a demonstration without giving students an opportunity to understand it or without utilizing all of its teaching possibilities. The teacher should never forget that even the common and simple ideas being developed are quite new to many students. Important points should be indicated, the relation to previous facts or principles stressed, and the implications of the demonstration pointed out. Sometimes it may be repeated profitably with students knowing what to expect. On occasion another similar demonstration may be performed, with different materials, or contrasting results.

Purposes Not Achieved Automatically

While the criterions suggested are valuable guides against which to check a demonstration, the meeting of these criterions is not an assurance of its effectiveness. The demonstration would probably be much less effective if it did not meet such criterions, but the purposes for which it is given are not achieved automatically. Planning is as necessary for student activity as for teacher activity. Such a demonstration as that of the bucket experiment frequently used to illustrate Archimedes' principle does not immediately and automatically illustrate to pupils the obvious generalization which the teacher so glibly reels off, that an "object immersed in a fluid is buoyed up by a force equal to the weight of the fluid displaced." It seems so clear to many teachers that it hardly needs mentioning: the water poured into the bucket, which has the same volume as the cylinder, has just the right weight to restore the balance lost because of the buoyant force on the cylinder.

Demonstrations Supplement Other Experiences

Excellent as demonstrations may be, they cannot be used as the only kind of experience. They should supplement other experiences and others should supplement them. Whatever the objective, demonstrations should be used along with textbook assignments, directed study, problem solving, laboratory experiments, visual and auditory materials, and trips. There is no one procedure or experience which can serve alone for all purposes and at all stages of understanding and achievement.

Student-performed Demonstrations

Students may give demonstrations before the class. While it is the unusual boy or girl who can show the skills and draw the conclusions the teacher may wish, there are occasions when it is desirable for students to give the demonstration. On occasion one or more students may set up, or help set up the apparatus. At other times they can assist the teacher in operating certain devices. Often it is important for the personal development of an individual or for his relationship to others in the class that he be given the opportunity to help in a significant way with a demonstration.

It should be remembered that demonstrations are usually given for the benefit of all the students rather than for the one or two who may participate in setting it up or giving it. Unless students can do as well as or better than the teacher in achieving the purposes, any extensive participation may be questioned. To the extent that student-performed demonstrations are monotonous, uninteresting, not clearly presented, promote levity, or tend to drag, they are probably not effective in promoting desirable outcomes. Unless students have been carefully prepared they are not likely to be able to make the same contribution as the teacher, who knows the needs of the group, the results of previous teaching, and the plans into which a particular demonstration is to fit.

Fixed Demonstration Tables

Giving demonstrations suggests the need for a place to give them. The usual arrangement in a room planned for science teaching is the fixed demonstration desk, which from many standpoints is the most satisfactory. It has the advantages of permanent connections for gas, water, and other facilities. The table should be about 36 in. high, 24 in. to 30 in. wide, and 36 in. to 48 in. long. The most desirable length depends on the size of the room and the use that is to be made of the table. The general structure of the table depends in part on the nature of the science courses to be served.

If the provisions for demonstrations are for only one kind of course, for example, chemistry, the variety of provisions may be limited. However, it has sometimes proved unwise to restrict the use which may be made of the table, not only because of subsequent needs for other courses, but also because of the enrichment of any given course through a wide provision of demonstration facilities. All things considered, the provision of general demonstration facilities seems justified in the usual high school.

The top of the demonstration table should be of acid proof material. The top may be of wood, stone, or plastic. If the top is of wood, it should be acid proofed. The facilities of the demonstration table should include hot and cold running water, a sink, AC and DC outlets, with provision for various voltages. They should include gas and, if possible, compressed air. The top of the demonstration table should include an auxiliary adjustable platform for two levels. This adjustable platform may be provided in various ways. A wooden box of sturdy construction serves quite satisfactorily. It should have dimensions of approximately 8 in. X 12 in. X 16 in. The auxiliary platform provides two levels for a variety of experiments, such as the use of the siphon, the inclined plane, and corresponding demonstrations.

It is desirable to provide for the control of obnoxious vapors on the demonstration table. One way of making this provision is by the installation of a commercial type of fume hood in the front portion of the room. Such a fume hood may serve not only for demonstrations but for individual student work as well. If it is well located and properly ventilated, those demonstrations in which chlorine, hydrogen chloride, and other obnoxious gases are produced may be very well controlled. It is possible to secure a commercial fume hood which is built on the demonstration table. Such a provision has the disadvantage of lack of flexibility. This may be objectionable in some schools, although certain of these individual fume hoods do not render a demonstration table entirely inflexible. An improvised fume hood utilizes a vacuum sweeper which is located in the basement, 1 although the location is not particularly important. A tube connects the chamber of the vacuum sweeper with the laboratory desk. At the surface of the desk a jointed tube is connected to a hood which is a glass half-bottle on a gooseneck. The electric control of the sweeper is installed on the demonstration table. When not in use the hood is removed and stored.

A demonstration table may be made more valuable through the provision of overhead projection facilities for lantern slides. The stage is a glass plate which is flush with the top of the demonstration table. The illumination is provided from beneath.

Plates may be installed on the surface of the demonstration table to receive vertical bars. Such plates may be so made that the bars screw into position, or the bars may be tapered in such a way that they are held rigidly without the use of threads.

Fixed Demonstration Table Not Essential

A good demonstration table with well-stocked cabinets and drawers, supplied with gas, water, electricity, and compressed air facilities encourages the giving of demonstrations. Such a table is not an absolute essential for the laboratory. It is possible to make use of an ordinary table of almost any type, including the usual office-type desk. Although the lack of service facilities is admittedly a handicap, even these can be improvised. Pails of water, jars, alcohol lamps, dry cells, storage batteries, extension cords, electric hot plates, and tires filled at a local service station are some of the provisions which determined and ingenious teachers have substituted for one or more of the usual service facilities.

Movable Tables for Demonstrations

In many situations the use of a movable demonstration table helps to meet certain limitations or difficulties. When the demonstration must be prepared in a free period in order to have it ready for a later period (or for the next day), the equipment may be partially or entirely arranged on the movable table. The table is stored in some convenient place until it is needed, then wheeled into the room in the few minutes before class starts. When the period closes, the table is wheeled again to a convenient point and the equipment and supplies disassembled at the convenience of the teacher. Preparations are thus easily made, equipment is protected, and limited space is conserved.

It is sometimes desirable to have the demonstration given from a position in the science room other than that at which the fixed demonstration desk is located. The lighting, nature of the activity, or arrangement of facilities may make this essential. Occasionally demonstrations are prepared in one room and are given in another. In all such cases, the movable table is of considerable assistance.

Nature of the Movable Table

The movable demonstration table may be equipped with heavy casters or rubber-tired wheels. These wheels should be 4 in. to 6 in. in diameter. It may be desirable to equip the legs of the table with doorstops so that it may be locked in one position. The movable table should be of the same height as the fixed table. The other dimensions may be somewhat less.

Running cold water may be provided on such a table by gravity feed from a bottle on a shelf. This shelf may be located at one end of the table or may be fastened on the wall of the room with rubber tubing leading to the demonstration table. If running water is available in the room, a tube may be connected to the outlet. A small sink may be installed on the table to receive waste. The sink may be connected with a large bottle or jar located in the lower part of the table.

An extension cord may be mounted on brackets on the rail on the underside of the table. The cord terminates in a double socket of the wall type attached to the rail. Such a socket facilitates electrical connections.

Ceiling Support

Demonstrations are facilitated in many instances by the provision of a ceiling support directly over the demonstration table. A support may be built so that it will be above the table and extend to one side. Such a support should be securely bolted to the structure of the ceiling and on its underside should provide, through projecting bolts with wing nuts, for the addition of various kinds of supports. Such provisions make possible such demonstrations as those with pendulums, pulleys, falling bodies, and wave apparatus.

An alternative to this method of support is the provision for a tall, rigid support from the floor or from the top of the demonstration table and braced against the ceiling. Such a support, however, has the disadvantage of interfering with demonstrations in some instances.

General Nature of the Demonstration Equipment

Certain general qualities of demonstration equipment should be kept in mind. Usually the equipment should be simple, without extra devices which may distract attention, and large enough to be seen easily from all parts of the room. The equipment should be clean so that it is always ready for use. It should be stored readily so that it will not stand where it may become soiled or misused.

The equipment for general use in demonstration should be stored as near the demonstration table as convenient. This may include such basic equipment as burner, ring stand, ring clamp, wire gauze, glass tubing, balances, weights, beakers, meter sticks, thermometer, and the like. Certain large devices such as levers, pulleys, pendulums, and inclined planes may be included. A galvanometer is quite desirable in some instances. If possible, it should be mounted on the wall or on a sturdy pier in the front of the room. The leads to this galvanometer should be either on the demonstration desk or should be readily accessible. The pier may also support large balances.

Lighting for Demonstration

Care should be used that the light from the window falls on the demonstration and not in the eyes of the students. The use of a rotating table, sometimes called a "lazy Susan," makes possible the changing of the position of the demonstration equipment so that all students may see. Certain demonstrations may not lend themselves readily to the use of this device, but it has been found to be quite helpful.

The science teacher should not rely on the use of the light from the out of doors nor on the general illumination provided in the room. Additional artificial illumination may be quite essential. The installation of an overhead floodlight or a spotlight at one side may be of considerable assistance. Such lights should be shielded so that light does not fall in the eyes of the students.

A rather simple and effective provision for additional illumination is in the form of the ordinary desk lamp which is adjustable. This may be readily placed for the best illumination possible. A somewhat more elaborate provision for illumination consists of a counter-balanced tubular bulb which is shielded. This bulb may be adjusted so that it does not interfere with the use of the blackboard in the rear. All accessory equipment which helps to improve the illumination should be so stored that it is readily available. The controls for the lights should be on the demonstration table or near at hand.

Sources of Demonstrations

Most high-school science textbooks do not have very extensive and specific suggestions for setting up and carrying out demonstrations. Science teachers should exhibit some resourcefulness in devising this type of student experience. There are, however, some excellent sources of demonstrations prepared and tried out by others, with which science teachers should be familiar. Later sections of the present treatise are devoted to specific demonstrations, laboratory work, and projects. Other sources where specific demonstrations are described are: professional magazines, such as School Science and Mathematics and The Science Teacher; science magazines written for the layman, such as Popular Science Monthly; trade journals, such as Chemist Analyst; textbooks and laboratory manuals; and special books on project and demonstrations.

Certain experiments ordinarily used as individual laboratory work may be given sometimes before the group as demonstrations. In some instances this is to be preferred, because the procedure is unfamiliar, the concepts are difficult, or the skills are new. In other situations the experiment calls for an accuracy of performance beyond the ability of the students, or the time favor is invoked, or the amount of equipment is limited.

It should be clear that students do not have the same kind and amount of experience in a demonstration as in an individual laboratory experiment. They get a different experience which under the circumstances seems to be more appropriate or valuable than the other experience would be. This suggests specifically the recurring question concerning the relative value of demonstrations and individual laboratory work.

▪ *Laboratory vs. Demonstrations*

Much has been written about and some research has been carried on relative to the merits of the laboratory as compared to the demonstration. What has really been proved is questionable. The demonstration is not to be thought of as a possible substitute for individual laboratory work. From the point of view presented here, one provides the kind of experience--in this case laboratory or demonstrations--which is the most appropriate for achieving the outcome desired.

This is considered in the light of the students' needs, their past experiences, and the plans for experiences ahead. Both should be used: at times, one; at other times, the other. Because of the individuality of students, with their differences in ways of learning, there is no one technique or procedure which can and should be used to the exclusion of others. Because the laboratory and the demonstration are alike in certain important respects, they are not to be considered as competing experiences, any more than motion pictures compete with slide films. Little evidence concerning the relative effectiveness of demonstrations and individual laboratory work can be forthcoming unless progress can be accurately measured. It is questionable whether short tests of subject matter measure the values to be achieved from the use of either of these procedures.

Individual Laboratory Work

The use of the individual laboratory in high-school science classes assumes that there are individual needs, and that there should be provision for individual growth through the laboratory approach. As the so called individual laboratory has been used in science teaching in secondary schools, the term is a misnomer. There are probably few schools in which students actually work entirely on an individual basis. A common pattern of work provides for students to work in pairs.

There are reasons why students should work in pairs, and even in larger groups, in the informal atmosphere of a laboratory. Such reasons are social and psychological in character. Attitudes and habits of cooperation, as well as willingness to exchange and weigh ideas, are encouraged by the type of laboratory experience in which the students are associated with others. In addition to such a socializing experience, there are adequate reasons for an entirely individual approach in the laboratory. A student should have experience in taking entire responsibility for the exploration of a problem and for drawing conclusions. Habits of self-reliance grow out of such situations but may be entirely neglected if a student is permitted always to work with other persons.

Use of the Laboratory Manual

Science teaching has come to rely very largely on the use of a laboratory manual. Many of the laboratory manuals have been prepared by competent teachers who, with a sincere desire to help other teachers, have prepared materials of great value if carefully used. Too often even the better laboratory manuals have had the effect of causing the laboratory work to become quite stereotyped. The goals too frequently are those of having all the blanks in the laboratory manual correctly filled and of completing as many of the experiments in the manual within the year as possible.

In spite of the good intentions of those who have prepared the manuals, the effect of the use of the manual has been to create a chasm between the student and the teacher. The teacher with his heavy load of work has assumed that if the student completes the blanks in the manual, he will have learned the associated content and will have developed attitudes and skills as well. Rationalization of the process has often been based on the argument of meeting the common needs of students through the use of common materials. However, in such stereotyped patterns of work, the usual effect is to prevent the student from having a true experience in the laboratory. The fact that neither the student nor the teacher had any part in planning the materials to be used and the procedures to be followed results in a rote performance which endangers the success of the efforts of both.

If individual needs are to be met through laboratory experience, dependence cannot be placed upon a set pattern of laboratory experiments. The-laboratory manual itself can be used, but not as a source of readymade procedure, the basis for which is clear probably only to the teacher.

The Laboratory for Experimentation

Experimentation can occur in the laboratory only to the extent that students work in terms of problems which they feel. It does not matter that the problems which they attempt to solve have been problems which students before them have worked upon and problems the answers to which the teacher knows. The important point is that the student work on a problem the answer to which he does not know.

Students can approach such problems intelligently as they participate in isolating and refining problems and planning for their solution. The implications of such planning include provisions for different problems for different students, different procedures to be used with different problems, and a wealth of resources to be used in the problem solution. The sources of problems for laboratory investigation may be found in a variety of activities of the individual and of the group. The problems may arise in class discussions or in project work. They may arise as a result of the student's school experience outside his work in science. They may come up in the student's reading or as a result of previous laboratory investigation. The sensitive teacher will be aware of those problems having implications for laboratory investigation and may make mention of the fact at the time that they arise, with an indication that they may be profitably investigated in the laboratory.

Uses of the Individual Laboratory

The use of individual laboratory procedure may serve any one of several purposes. Two purposes that have frequently been associated with the typical laboratory procedure are (1) development of skills and techniques, and (2) verification of principles or phenomena already discussed. As objectives of individual laboratory work, both are worthy. There is serious doubt as to whether these should be the only objectives served and whether they are actually achieved by the usual procedures. The lack of satisfaction of science teachers with the results secured in their teaching poses a serious doubt as to whether they should rely upon the laboratory for verification and confirmation. There is doubt, also, as to whether experimental methods or critical thinking can be effectively taught in the absence of real experimentation on the part of the student. Undoubtedly the student needs experiences in developing skills, in the reading of instruments, in the acquiring of certain attitudes relating to reliance upon the results of an experiment, but it is quite questionable whether he learns to make the necessary predictions and hypotheses or to draw valid conclusions without actually having occasion to do so. There seems to be no substitute for the actual situation in teaching for methods of science. There is adequate reason to believe that the slavish following of directions with the repetition of steps in a process not only is a waste of time but serves to rob the student of initiative.

Sources of Procedures

The substitution of plans and procedures formulated by the student with the guidance of the teacher for a stereotyped procedure is a challenge to the science teacher. He must be able not only to assist students in isolating and formulating their problems but also to provide suggestions for initial steps toward their solution. Such initial steps include the direction of the student toward sources of information which are likely to be of assistance.

At such a point as this the laboratory manual, along with other sources of techniques and procedures, becomes an appropriate resource. There should be a supply of manuals of different kinds. In addition, there should be such sources as formularies, books describing projects and techniques, and such periodicals as Popular Science Monthly. From such sources the students may secure a variety of ideas as to appropriate methods of solving the problems which they have found. The teacher must be sensitive to the fact that time is necessary for planning the solution of problems. Students not only should have access to such sources as those suggested but should have class time to use them. Each student should prepare proposed procedures for his activity. In conference with the teacher these can be elaborated, particular points emphasized, and cautions of difficulty or danger introduced.

It is clear that in the use of such developmental procedures for laboratory work the usual number and kind of laboratory "experiments" cannot be included in the work of the year. The value of such developmental procedures lies in the quality of the work accomplished rather than in the quantity; the advantage becomes more real as the teacher reflects upon the ineffectiveness of the typical following of directions which has characterized so much of the use of the laboratory.

In some institutions situations the teacher may be obligated to use one laboratory manual, and only one. The difficulties arising from this use may be mitigated somewhat through exploration with the students of the plan for laboratory work as set forth in the manual. Such discussion and planning may include a clarification of the meaning of the directions in the manual, the basis for choice of techniques and materials, and the reason a given procedure in the manual has been chosen to assist in the solution of a certain problem. Likewise, this exploration may assist in making the student sensitive to further problems that may grow out of his laboratory work, as well as to suggest to him some independent activities which he may wish to follow up in the laboratory or in his free time. The teacher should be able to supplement the directions and suggestions in the laboratory manual from his own background in answering questions, providing additional directions, and clarifying obscure points. The laboratory can promote the location and solution of problems if resources are available. If student interests and needs are met, the range of problems is broad; the resources necessary for individual exploration are correspondingly numerous.

Laboratory Arrangement

The arrangement of certain facilities of the laboratory is, in general, rather static. Plumbing and electrical connections require that tables remain in fixed position; changing the position of such facilities is a major operation. Alteration of such tables should not be considered an insuperable task, but the planning for it should be done with care and with thought for its function.

Auxiliary tables may be secured to make the arrangement of the laboratory more flexible. The tops of such tables may vary in height, according to the purposes which they are to serve. Such tables can be moved from one position to another as their use over a period of time varies. These auxiliary tables are preferably somewhat small but should be quite sturdy. If they are equipped with drawers, a certain amount of material and equipment can be stored for ready use.

Organization of Laboratory Facilities

In many schools, particularly in those that are larger, there are separate laboratories for different subjects--chemistry, physics, general science, and biology. In the moderate-sized and small high school it is not uncommon that two or more areas of science are taught in the same laboratory. Where a laboratory is to serve a single subject area, such as chemistry, it is necessary that primary provision be made for the techniques and procedures of that field. In spite of such specialization there should be at least demonstration facilities of a quite general nature not only because of the desirability of being able to use the room for classes in other fields but also because of the enrichment which can be provided in any particular field by calling upon the content of another field.

In the moderate-sized and smaller school and in the general science laboratory of the larger school, the need for a wide variety of facilities within a single laboratory is even more imperative. In those cases where the laboratory serves not only the chemistry and physics classes but the general science classes as well, it is essential to make provision for the specialized fields as well as the more general ones. For example, there should be adequate equipment and supplies for many students to have experience with electricity at the same time. Similar provisions are necessary in other areas where a number of students are to be served. In addition to such provisions of a general nature, there should be facilities to serve the needs of the general science classes.

Such needs have been met on occasion by the provision of certain functioning areas which may serve both the general and special sciences. These areas or centers are chosen on the basis of a particular field of concern or a particular technique; for example, a certain portion of the laboratory, perhaps a desk or table in one corner, serves as an electrical center. On and around it are electric outlets, a control panel, and equipment and supplies of electrical nature. In another part of the room a center on air pressure is set up. Such a center has a pressure and a vacuum pump, bell jars, and various pieces of equipment which may depend upon air pressure for their operation. Other such centers of activity may include analytical chemistry, weather, aeronautics, and the like.

The choice of such centers should depend upon the needs of the local situation. Such centers have the effect of grouping together equipment needed for a particular type of activity, as well as distributing students over the laboratory in terms of their needs. It is necessary on occasion to move equipment from one center to another, or to move the equipment from a center to a more generalized portion of the laboratory. However, such centers have the effect of providing for more efficient organization of work when project activities are under way.

It is desirable in many cases to provide for certain other kinds of centers. In some instances it may be helpful to provide for a reading center. At such a point all reference materials, including the file of bulletin board materials, are located. If another room is available, such a reading center may profitably be transferred. A shop center may be necessary in some laboratories. In other instances such a center with facilities which have been previously outlined may be located in a different room. Quite often it is desirable to provide a center for visual materials, including the filmstrips, bulletin-board materials, and the like. Such materials may be grouped with the reading materials if desirable. Freedom to move from one center to another and to use all the available resources under supervision tends to encourage an approach which will lead to true investigation and problem solution.

Adaptation of Existing Facilities

The science teacher may often find that changes in facilities must come about slowly. However, through such major operations as are necessary in rearrangement and through minor adaptations, many of which can be carried out through the teacher's efforts and through the efforts of students, it is possible to organize most laboratories into functional patterns which will provide for and encourage learning activities for the students.

CHAPTER 5

IMPROVEMENT IN LABORATORY OUTCOME OF TEACHING

Among the objectives of the modern secondary school those relating to teaching for reflective thinking have recently received considerable emphasis. Many advocates have stressed the need for improvement in thinking as being more important than achievement in subject matter. The importance and desirability of planning for experiences concerned with thinking have been widely presented in reports, courses of studies, and books. 1 There is general agreement that in a democracy, where citizens are expected to participate in helping to solve community and national as well as personal problems, everyone should be able to think critically about problems and their solution.

While each of the major areas in the secondary-school curriculum offers many and varied possibilities for providing experiences in thinking, none is more replete with potentialities than science. Here the very terms used suggest various aspects of thinking--scientific method, experiment, control, test, verification, conclusion, authority.

Thinking Not Automatically Achieved

Science teachers, however, do not need technically devised tests to convince them those students who have been in the usual science course, even in physics or chemistry, do not necessarily and automatically emerge with a desirable proficiency in scientific or reflective thinking. It is not difficult to recall numerous examples of scientists, recognized as experts or authorities in their special fields, who have not exhibited their scientific thinking and attitudes in other problem situations such as those of social, economic, or political character. To work in a science laboratory does not guarantee generalized outcomes of reflective thinking, though it may have excellent possibilities of helping to make progress toward this goal.

In the discussion of the use of the laboratory it was pointed out that many student experiences in the science laboratory may be quite perfunctory and devoid of thinking. Some laboratory work may even engender outcomes which are antagonistic to scientific thinking because of the procedures and emphases. The blind following of directions and filling in of blanks for reports is not likely to be good experience in critical thinking. A demonstration performed by a teacher who points out what is happening and indicates the conclusion which should be drawn or how it illustrates a particular principle may furnish little experience in thinking. This limitation need not exist.

Laboratory Work as Experiences in Thinking

Any work with apparatus and materials such as demonstrations, laboratory work, or projects has some obvious possibilities for experience in thinking. Here, while recall is important, something more than memory of facts or statements of principles from the printed page should be developed. A wide variety of processes relating directly or indirectly to critical thinking comes into play with different kinds of work with apparatus.

Making accurate observation, devising apparatus, predicting, remembering, applying facts and principles to new situations, interpreting observations, formulating and testing hypotheses, finding the reasons for results, planning controls, withholding judgments, being tolerant these are some of the experiences which students may have. However, students are likely to have these only when the teacher, or the teacher and the students together, plans to provide such experiences. Before suggesting how such experiences may be provided, the nature of thinking should be considered.

What Is Thinking and Teaching for Thinking?

While there are, from some standpoints, some important differences in such terms as reflective thinking, critical thinking, scientific method, reasoning, and problem solving, for the present purposes the phrase reflective thinking, or simply thinking, is used to indicate the achievable objective which seems to be suggested by these terms. As has been pointed out elsewhere, there is no one scientific method, or the scientific method. There are scientific methods, adapted to the nature of the problems, the field in which they fall, and the data and techniques with which they must deal. The methods which can be used for attacking problems in nuclear physics may be quite different from those used in problems in anthropology. Both may be judged as scientific, however, by competent workers in these fields. This suggests that there are no "steps" or "elements" which must be followed invariably in order that a problem be solvedú Problems vary in their nature and in their complexity. The steps or procedures which have to be carried out in the discovery of a new element or the invention of a complicated machine are at least more numerous and elaborate than those, for example, in using a table of wet- and dry-bulb thermometer readings to find relative humidity. The latter, however, is a problem to be solved, involving interpreting data presented in tabular form.

There are countless relatively simple problems involving critical thinking with which students are confronted dailyú. Many of these are similar to the problems with which adults have to deal. It may be helpful to make use of two analyses which have proved satisfactory in working with high-school students toward this objective of thinking. One analysis suggests that the process of reflective thinking may be analyzed into the following five phases:

1. A sense of perplexity, or of want, or of being thwarted, followed by identification of the problem. This may or may not be formulated in words.

2. "Occurrence" of tentative hypotheses. These hypotheses may be based upon a supposed relationship between what is perplexing in the situation and what has been previously experienced.

3. Testing and elaborating hypotheses, by imaginary experiments, which may on some occasions be supplemented by recourse to paper and pencil experiments and on others to even more concrete practical experiments. This process may also require reference to books or other records.

4. Devising more and more rigorous tests to which the resulting hypotheses may be subjected. Testing the hypotheses in these situations either solves the problem or reveals that the hypotheses do not stand the test of action.

5. Arriving at a satisfactory solution and acting upon it. This may include devising a form of statement by means of which the conclusions may be expressed and communicated with the highest possible precision.

The reference cited points out that no one of these phases is really distinct and discrete but that all may be involved "more or less simultaneously," each supplementing the others, but that an attempt to identify the abilities and attitudes involved in them may aid in working out ways of teaching as well as testing which may make progress toward contributing to the ability to think reflectively. Accordingly, the following general and special abilities are listed.

General Abilities

A. To recognize clearly (in imagination, with or without statement in communicable form) what the problem is

B. To bring past experience to bear effectively in making preliminary guesses as to what is the crux of the problem, that is, in guessing upon what factor in the situation it is most efficient to concentrate

C. To recognize what things previously learned (whether from experience or books, or both) may contribute to the solution of the problem

D. On this basis to form hypotheses that will serve as a basis for imaginary or practical experiment

E. To marshal evidence which either supports or invalidates any given hypothesis?

F. To see (though not necessarily through any formal logical process) what the crucial tests to which are an hypothesis can be subjected

G. To devise experimental conditions or to find analogous situations which will test the hypothesis most conclusively?

H. To express conclusions unambiguously

Special Abilities and Skills

A. Skill in devising and setting up experimental apparatus

B. Skill in the use of measuring instruments

C. Skill in recognizing sources of error in observation and measurement

D. Skill in getting the sense of a written passage

E. Skill in using libraries and other sources of graphic information

F. Skill in assessing the reliability of authorities

G. Skill in expressing hypotheses and tentative conclusions unambiguously and economically in words or other symbols

H. Skill in mathematical procedures, including statistical methods and graphing

I. Skill in conducting discussion in ways to bring the main issues to the forefront

Attitudes or Dispositions

A. Active curiosity

B. Caution in making generalizations and readiness to revise them in the light of new evidence

C. Tolerance of new ideas and suggestions from all sources

D. Disposition to search out and try out a variety of approaches and points of view on a problem

E. Confidence that scientific methods will be successful in solving problems

F. Disposition to see a problem through to its conclusion in spite of distractions

G. Readiness to act on the basis of tentative judgments.

Make Up and Carefully Plan an Experiment to Find Out Whether the Answer You Selected Is the Right One.

In this case you would plan an experiment to find out whether the water on the potato causes the spattering.

Plan to do an experiment in two parts whenever possible.

In this case one part would be to fry a slice of potato that has water on it, and the other part would be to fry another slice like the first one except that you wipe all the water off it before you fry it.

These two parts of the experiment are exactly alike in every way except one, and that is called the Experimental Factor. The only difference between your two experiments is that one slice has water on it and the other has not. Water Here Is the Experimental Factor, because it is the one factor that is different in the two parts of the experiment.

The part of the experiment that does not have the experimental factor in it is called the Control Experiment.

As a control, you will fry some slices of potato in hot fat Without Any Water on Them. This can be done if the slices are carefully dried with a towel before being dropped into the grease. The part of the experiment which includes the experimental factor may be called the True Experiment.

As the true experiment, you will fry some slices of potato in hot fat With Plenty of Water on Them. This can be done if the slices are taken directly from a bowl of water and dropped into the grease.

Remember that the experimental factor must be the only difference between the true and the control experiments. Then any Differences in the results of the two experiments will be due to the experimental factor. Carry Out the Experiment with Great Care According to the Plan. Make as careful observations as you can. Whenever your answer must be exact, make careful measurements of the results if you can. When the two experiments are carried out, you find that the dry slices only sizzle, while the wet slices cause the grease to pop and spatter.

Repeat the Experiment to See whether You Get the Same Results the Second Time. This Second Experiment Is Called a Check Experiment. Sometimes, instead of repeating the same experiment, an entirely new experiment using the same experimental factor is carried out as a check on the results.

Since water appears to be the cause of the spattering, you might take a few drops on your fingertips and let them drip into the hot grease, to see whether water alone behaves in the same way.

Your conclusion should be stated so that it indicates whatever your experiments show, and only what they show. For example, your conclusion might be: "It is the water on the sliced potato that makes the hot fat spatter from the pan." Use the Facts You Have Thus Learned When You Face a New Problem That Is Similar or Related to This One.

As a result of this experiment, you might decide to dry the slices of raw potato upon a towel hereafter, before dropping them into the pan, in order to avoid so much spattering of the grease.

Perhaps you might carry your investigations further by trying to find out whether it is the water in other foods, such as raw eggs or meats, which causes various cooking fats to spatter. It is by using this method over and over again that our knowledge of the world about us increases; and the more we know about the world around us, the better we are able to live in it and to control it. While, as has been pointed out, 8 students' problems are not necessarily the same as the teacher's problems, the above analyses have proved helpful for both the teacher and students in arranging problems that are challenging and interesting and involve aspects of thinking. Some specific ways in which demonstrations, laboratory experiments, and projects can be used for one or more general or specific abilities involved in thinking are presented for illustrative purposes.

Using Demonstrations for Teaching for Thinking

Teaching for thinking is largely a matter of utilizing the usual teaching procedures appropriately as the teacher goes about teaching scientific concepts, facts, and principles. This objective cannot be achieved through a unit on scientific method somewhere in the general science, biology, chemistry, or physics course, followed by an emphasis on the "facts" during the remainder of the year. If a mastery of facts is an important course outcome, its realization may be furthered by appropriate experiences in thinking.

A particular exercise or experience in thinking may or may not be appropriate to use in a particular situation. A certain demonstration for one class may be given to help teach a particular fact of science, at another time as an experience in accurate observation, at another to utilize previous knowledge for prediction, at still another as an application of a recently studied principle. The following examples have been used in physics, chemistry, or physical science classes with tenth-, eleventh-, or twelfth-grade students. They are illustrative of one satisfactory way of using demonstrations to emphasize thinking.

Use of the Laboratory in Teaching for Thinking

Certain of the difficulties in the use of the laboratory in the secondary school have been pointed out. The general technique there proposed obviously aids in achieving an outcome such as reflective thinking. If students are allowed to choose and set tip their own laboratory problems and to participate in planning the procedures for solving certain proposed problems, there is abundant opportunity for thinking. The clear definition of the problem and statement of the hypothesis to be tested are very important phases of thinking, and experience too often foreign to students' laboratory work in the secondary school.

The use of "controls," to mention just one phase of procedure in laboratory work, is a concept and a technique with which high-school students are often unfamiliar, essential as it is to the scientific method. Where students are encouraged to set tip and try out problems involving controls, these do provide worth-while experiences in thinking. And since "the value of problem-solving through laboratory work in the school does not lie in the factual knowledge that may result from it but in the attitudes and habits of reflective thinking it encourages and in the understanding it gives of how the knowledge of science gained by the student from description was attained in the first place," all such experiences relating to thinking are desirable.

The use of the individual laboratory makes provision for situations conducive to reflective thinking. Such situations are primarily problematic in nature; being problematic, their origin must be in the student's own experience. When the student is required to adopt a problem, such as the question at the top of a sheet of laboratory exercises, or a statement by the teacher, the significance of the required problem is lost and the real problem, the one which the student feels, is that of meeting a required assignment. Because he may feel no genuine concern in relation to the adopted problem his efforts are likely to be perfunctory. The use of laboratory experiences when they are needed, the basing of experiences upon problems felt by the students, the participation of the students in defining problems and determining procedures, and reporting appropriate to the problem attempted and conclusions drawn, give a range of opportunities for experience in thinking.

Few studies have been made concerning the method of laboratory work most conducive to scientific method on the high-school level. One by Thelen, relating to general chemistry in college, is suggestive. On the basis of results obtained from a study of different methods of teaching for scientific thinking, the following "Suggestions for Teaching Chemistry" were made, along with others. While these are directed to college chemistry, many may apply to the teaching of high-school chemistry, as well.

Suggestions for Teaching Plan of Practical Chemistry

1. Insofar as possible, plan activities requiring group planning and execution, with responsibility divided among individuals.

2. Develop experiments in class discussion. The saving of time in the laboratory would be considerable if twenty minutes of the quiz hour could be devoted to developing the rationale and technique of each experiment.

3. Follow up each experiment in class discussion, formulating as a class the best possible statements of conclusions, and identifying the assumptions and sources of experimental error in the experiment.

4. Use the development of chemistry as the basis for developing understanding of scientific method, and provide real opportunity for the student to use scientific method in planning and following up experiments.

5. Eliminate or redesign the test-tube survey type of experiment, except when it can be shown that the range of superficial experiences with a wide variety of chemicals results in valid and important generalizations.

6. Plan each experiment (except those whose function is exploratory) to involve prediction of results and comparison of results with prediction.

Providing for a Variety of Student Problems

When students work upon different problems, the seemingly inevitable result is a confusion which may be disturbing to the beginning teacher. The need of students for a variety of equipment and materials and for advice on a wide range of problems is sometimes perplexing but can be made challenging and fruitful for everyone concerned. As students and the teacher counsel together, the quality of the learning situation is improved. The sensitive teacher recognizes that the confusion diminishes as the students progress in the solution of their problems, and the confusion is really a characteristic of this type of learning situation. As the work progresses from day to day the activities take on more order. Students approach new problems at different times and the demands on the teacher are correspondingly distributed. A factor that tends to reduce the confusion is an adequate supply of physical resources, which should include the supplies and devices needed for the usual laboratory work, and others as well. The exact scope of such supplies can perhaps never be forecast for the high-school laboratory.

Arranging Laboratory Facilities

In using unconventional laboratory problems, students have need for devising and constructing apparatus to fit the problem. A small array of tools, which may be of the home workshop type, is quite desirable. Such power tools as a drill and a circular saw, a jig saw, and a joiner are suggested. A small wood-turning lathe and a metal lathe are useful on occasion; if these can be used in the industrial-arts shop, it may be unnecessary to have them in the science laboratory. Hand tools also are needed for this kind of laboratory work. A hand drill, vise, brace and bits, various saws, pliers, wrenches, and similar common tools should be in the laboratory or easily available to students in an adjoining room. The thinking involved includes in part the planning and decisions concerned with the devices to be made and the actual making of them. With experience in this kind of laboratory procedure, students rely less and less upon the teacher. They do more of their own thinking.

CHAPTER 6
USAGE OF AUDIO VISUAL MATERIAL

One of the problems that confront science teachers is making the learning functional. Much of the content of science is relatively abstract and therefore difficult to comprehend. There seems to be no single an to this perplexing problem, but one of the approaches which has been found to be of major assistance is the use of visual and auditory materials.

Many of the materials and procedures are not unique to the teaching of science, but their contributions to learning in this field are so significant that the science teacher has an obligation to profit by the teaching opportunities presented. Of the variety of materials and procedures which have been classified as visual and auditory the following are considered in this part: the use of motion pictures, lantern slides, film slides, opaque materials, and microscopic materials; the use of the blackboard and such graphic materials as maps, charts, and graphs; the use of the radio, the recording, the public-address system; and certain miscellaneous materials of value to the area of science.

The Motion Picture Usefulness

The contribution of the motion picture to the learning situation in science may be summarized as the providing of greater reality. The separate phases which contribute to this reality may be classified as detail, motion, continuity or process, and unobservable action. In addition, there may be sound and color. It is recognized by the teacher of science that these separate phases exhibit their values in such ways as the portraying of motion where the operation of machinery is being studied; in giving a sense of continuity which is essential to the understanding of certain processes, and to cause-and-effect relationships; in suggesting the nature of unobservable action in such situations as the activity within opaque chambers and channels, and the flow of electricity.

There is an abundance of films of value to the science teacher. These films are provided by educational organizations, commercial producers and distributors, industrial organizations, and by national and state governments. The science teacher need consider only the suggestive titles of films to appreciate the range of available resources. In the course of his teaching he should build a card file of films as a part of his physical resources for instruction. With the increasing availability of projectors, the science teacher should become familiar with the operation of the projector, the sources of films, the nature of the films, and the provisions in his particular school for the use of motion pictures.

Operation of the Motion-picture Projector

The care and operation of the projector depend in part on the type of projector which is used. There are certain essentials of care and skills in operation which pertain to all projectors.

Lantern Slides and Projectors

The use of the lantern-slide projector commends itself to the science teacher because of the wealth and variety of resources which it provides. These resources include (1) commercially produced materials; (2) handmade slides, the abundance of which depends only upon the ingenuity of the teacher; (3) projected materials in the form of slides; (4) other projected materials and devices.

In some instances the lantern-slide projector has become such a traditional resource that its advantages are not entirely appreciated. These advantages include the following: (1) The time factor is readily controlled; a slide may be shown at the time desired and for as long a time as it is needed. (2) A single slide or small numbers of slides may be readily chosen and used. It is not physically difficult to separate those which are needed from those which are not. (3) Slides may be produced by the science teacher to meet his needs. (4) The equipment necessary for projection is generally available. There is probably only a small percentage of schools which do not have lantern-slide projectors. (5) The projection equipment is easily manipulated.

The use of standard lantern slides, 3¼ in. × 4 in., makes available a wide variety of illustrative materials. Many of these have been commercially produced. They include pictures, diagrams, and other graphic materials which are of value to the science teacher. Sets have been prepared to cover the various subdivisions in the fields of physics, chemistry, geology, weather, and general science. Such materials may be used as illustrative material in developing a unit of study, to introduce a unit, or as a basis for review. Manuals to accompany these sets of slides are supplied in most instances. The science teacher should not feel that an entire set must be used. He should familiarize himself with each of the slides and with the related descriptive material in the manual. Those slides which serve the teaching situation best should then be selected and used. Certain situations may be served best by the use of the slides with the entire class; in other situations the slides may be used best by one member of the class or by a few students.

The manual should not be regarded as something to be followed literally, but rather as resource material to make the slides more valuable. Often the teacher is able to use a slide to greater advantage if he does not read to the students the narrative supplied by the manual but draws on it only to supplement his own background.

A type of lantern slide with which the science teacher should be familiar is the so-called 2-in. × 2-in. slide. Such slides require a special projector or an adapter for the standard lantern-slide projector. The slides are made photographically on plates or from the pictures taken on 35-mm film. The pictures are separated and mounted to be used as lantern slides. As yet there is not the abundance of commercially prepared material of this size that there is in the larger size of lantern slide. The science teacher who has access to a camera which takes pictures on 35-mm film may find this method of enriching his resources a feasible and economical approach.

Use of the Overhead Projector

There are two general types of lantern-slide projector, the horizontal type being the more common. The overhead lantern-slide projector, while less commonly used, offers certain advantages over the horizontal projector. The advantages of the overhead projector are: (1) It is used on the demonstration or teacher's desk in the front of the room rather than at the center or rear of the room. (2) The teacher faces the students. (3) It is easily put into operation. (4) The slides are placed on it as one would read them. (5) Materials on the slide are easily referred to with a pointer. (6) Certain materials other than slides may be projected. Such projectors are available from companies handling visual equipment as well as from certain companies distributing scientific equipment.

Constructing an Overhead Projector

The science teacher can build a projector which serves quite satisfactorily. A 500-w projection bulb is housed in one end of a metal box, 6 in. × 6 in. × 12 in. (the dimensions are approximate). The metal is No. 22 or No. 24 iron. The lamp house end has ½-in. holes drilled near the bottom edge and near the top edge (at least six holes on either side) to provide ventilation. To prevent the escape of direct light from these holes, metal shields are bolted to the inside at a distance of about ½ in. from the wall. The base may be a wooden block to which the metal box is screwed. A concave reflector is fastened so that its principal axis coincides with the optical axis of the bulb-mirror-lens system. Midway in the box is a vertical metal partition which supports a piano-convex condensing lens.

The condensing lenses should be approximately 5 in. in diameter, the opening in the partition being slightly smaller. Metal clips hold the lens to the partition. The second condensing lens is placed as indicated with the plane side up, just under the top of the box, held there by metal clips. The opening in the top is that of the projection area of the standard lantern slide which is placed with its long dimension at a right angle to the long dimension of the box. It is located with its geometric center coinciding with the optical axis of the system. A mirror is placed below the opening at an angle of 45 deg with respect to the optical axis. For best results provision should be made for adjustment of the mirror.

The projection lens and second mirror are supported by a rod of ⅜-in. or ½-in. diameter which is bolted through the wooden base and braced against the metal box to secure rigidity. The height of this support is determined by the focal length of the projection lens and the dimensions of the second mirror. A height of 24 in. to 28 in. is satisfactory for the lens and mirror to be described. The support should be placed opposite the center of the second condensing lens. The projection lens should be of good quality, with a focal length of 6 in. or 7 in. and a diameter of approximately 2 in. The lens should be mounted, then secured by a clamp, such as a condenser clamp; or a burette clamp may have the jaws removed and the projecting threaded rod bolted through the wall of the lens mounting. The projection lens should be placed concentrically with respect to the second condensing lens.

Above this lens is a good mirror (an inferior mirror at this point seriously impairs the results). This mirror should be approximately 4 in. × 6 in. A clamp supports the mirror by its edges; it is adjustable in order to vary the position of the image on the screen. A ¼-in, bolt passes through the clamp on the mirror. A wing nut on this bolt facilitates adjustment of the mirror. The bolt (about 5 in. long) is secured by means of a universal clamp to the support rod. The uncovered surface of the mirror is placed toward the projection lens so that the rays are reflected to the screen. The back of the mirror is covered for protection.

The slide to be used is placed on the stage (S). The projection lens is raised or lowered until the image is in focus on the screen. If the screen is appreciably higher than the projection mirror, the screen may be tilted forward at the top until the image is not distorted. For a given room this position is easily determined, and the screen may be permanently mounted. A rigid screen supported by a frame serves best for this purpose. A screen distance of 4 ft to 12 ft is satisfactory, depending on the projection lens used and the desired size of the image.

Materials to Be Projected

There is a wide variety of standard lantern slides of dimensions 3¼ in. × 4 in. which the science teacher may use. In addition to these lantern slides, there are many variations in projection techniques which are of value in science teaching. A device of value to the teacher is the so-called hinged, or book, slide. This consists of two lantern slides hinged together along one edge, gummed tape serving as the hinge. Various types of materials may be placed between them and projected.

Demonstration Equipment for Projection

A projector may be used to show many actual experiments. In general, the overhead projector is more adaptable to this use than the horizontal projector. In the following suggestions the equipment described may be used effectively with the overhead projector, although in some instances adaptations may be made for the horizontal projector.

Wave Apparatus

A wave apparatus may consist of a helical coil of wire in a frame. By turning the handle at the point H the projection is placed on the screen, suggesting the motion of the wave. A bit of wax placed on the wire indicates the motion of the energy field, or it may be used to illustrate simple harmonic motion. A sheet of brass of No. 20 gage is of desirable quality and thickness for the frame, although other types of sheet metal may be used.

Electrolytic Cell

A petri dish or other shallow glass vessel may be used as an electrolytic cell. The dish is placed on the stage of the projector with a small amount

Wave apparatus. of electrolyte covering the bottom. The electrodes are placed in opposite positions in the cell. Each electrode is shaped like the letter T, the crossbar bent at a right angle to the stem. The crossbar rests in the liquid, the stem with the wire carrying the current resting on the edge of the dish. The rate of liberation of gases and the process of electroplating metals are readily demonstrated.

Polarized Light

Various demonstrations of the use of polarized light may be carried out. For this use, the projection lens is removed. Polaroid disks may be held in the beam of light above the stage by clamps appropriately arranged. Sheets of mica, plastic models, and cellophane may be placed in this beam of polarized light and the effects studied on the screen. Crystals of benzoic acid, aspirin, hydroquinone, potassium nitrate, borax, and potassium chlorate may be studied. These crystals should be produced by the slow evaporation of their solution upon a clear lantern slide.

The effect of a sugar solution on the plane of polarization of the polarized light is shown in a corresponding manner. For this purpose a polaroid disk is placed just above the stage. The sugar solution is placed in a glass cylinder, such as a hydrometer jar or graduated cylinder. This cylinder is placed in the path of the polarized light. The second polaroid disk (analyzer) may be placed above the cylinder or in a vertical position between the mirror and screen. In use, the analyzer is turned to the point of maximum extinction without the column of sugar solution. The cylinder containing the solution is then placed in the beam and the analyzer adjusted until maximum extinction is again secured. The rotation of the plane of polarized light is thus qualitative. If desired, the quantitative relationship may be carried out by marking the positions of the analyzer.

Kinetic Theory

The kinetic theory of matter may be demonstrated by placing upon the stage of the projector a plate of glass of dimensions approximately

8 in. × 8 in. A frame of wood of external dimensions of 6 in. × 6 in. is placed on the glass. Steel ball bearings of 3/16-in. diameter are placed inside the frame. (The diameter is not critical.)

Twenty-five to fifty of these ball bearings may be used. As the frame is moved in a rotary fashion the motion of the ball bearings is projected onto the screen and provides a two-dimensional representation of the motion of molecules. An analog to Brownian motion is demonstrated by placing steel ball bearings of two different diameters within the frame. Rotation of the frame demonstrates the effect of the smaller balls upon the larger.

Air-flow Cell

The projection of an air-flow cell is readily made. The cell consists of two plates of lantern-slide glass (3¼ in. × 4 in.) held by wooden pieces with a metal cell at the end. The metal cell has a series of baffle plates which serve to distribute the flow of the incoming gas. The gas is made visible by the use of a Dry Ice-hot-water generator. Small pieces of the Dry Ice are placed in the flask and hot water is added. The vapors are led by means of a rubber tube to the cell. They are distributed in the chamber at the end and flow evenly through the cell. Small objects in the form of squares, circles, streamlined objects, and wing sections are placed in the cell and their images focused on the screen.

Magnetic Field

Corresponding cells without the provision for the flow of air may be made to show various aspects of a magnetic field. For example, a solenoid may be placed within a cell. Iron filings are placed within the solenoid, and when a current is passed through it the nature of the magnetic field in the core is shown.

Magnetic Compass

A cell for showing the activity of a magnetic compass may be provided for the overhead projector. A compass needle of 1-in. to 2-in. length is mounted at its center point on a tack cemented to the surface of a lantern slide in a central position. The compass needle from a small encased compass may be used. The behavior of the compass when a magnetic pole is brought near is observed as it is projected on the screen. One pole of the compass may be identified by a very small tab of attached Scotch tape.

Electrical Instruments

Silhouettes of electrical instruments may be projected. By removing the metal back of the instrument, light is permitted to go through, showing the silhouette of the mechanism. For quantitative work the scale which has been removed may be reproduced on a piece of glass or celluloid.

Sky Charts

Lantern slides representing the sky may be made for use in either the horizontal or overhead types of projector. These slides may represent any section of the sky and are made by placing a paper mask between two lantern-slide cover glasses or pieces of glass. Holes are punched in the mask to represent the stars. These holes are placed in the relative positions of certain stars in a constellation. Larger holes provide for stars of greater magnitude. Star charts may be used as patterns. It is well to check the appearance of the constellation with the sky after the material is prepared. A more elaborate sky chart may be made by using circular pieces of glass with Polaris represented at the center of the glass. The stars in the northern sky may be represented in their proper positions. The two glass disks are then fastened together with tape. The disks may be rotated as the projection is made, thus representing the apparent motion of the stars in the sky.

Use of the Horizontal Lantern-slide Projector

The horizontal projector, using as it does a vertical cell, does not readily permit certain of the projections which are described in the preceding paragraphs. The projection of the lantern slide, book slide, and certain of the special slides is possible in the horizontal projector. Many of the demonstrations requiring special cells and other small pieces of equipment are not so readily accomplished with the horizontal projector. There are, however, some types of projection which seem to be better adapted to the horizontal than to the overhead projector.

The Projection Cell

For these projections a specially constructed cell is needed. Adaptations of this cell are made as they are needed. The cell is made to be inserted in the opening in which the lantern-slide carrier is normally placed. The following uses of the vertical projection cell are suggestive of the variety which the resourceful science teacher can readily prepare. Position and motion of the image on the screen are, of course, inverted.

Convection Cell

Convection is shown in a cell by placing a coil wound in the form of a helix. The coil is attached to the source of current with a rheostat in series. The rheostat is adjusted to the point at which the coil maintains a low red heat. When this adjustment has been made, the coil is permitted to cool and water is placed in the cell. The cell is placed in the lantern and the image of convection currents from the reheated coil is projected on the screen.

Relative Density

The relative densities of various liquids may be shown by placing in such a cell small quantities of mercury, saturated solutions of potassium carbonate in water, colored alcohol, and kerosene. The liquid is stirred and then projected. The observation of the separation into levels of various densities is an interesting and instructive experience.

Electrolytic Cell

By placing vertical electrodes in such a cell the electro deposition of various metals and gases may be demonstrated readily. Copper or silver plating and the decomposition of water into its components may be shown.

Capillary Action

Capillary action may be shown by placing the tubes of different diameters in the vertical cell containing a small amount of colored water. Plates which touch on one vertical edge and are nearly together at the other edge may be used instead of the tubes.

Surface Tension

Effect of surface tension may be shown by supporting a camel's-hair brush in water in the projection cell. As the brush is raised from the water the effect of the tension is noted on the screen.

Chemical Reactions

Certain kinds of chemical reactions can be carried out in a vertical projection cell if the space above the cell is open. This may be accomplished by removing a portion of the frame which supports the slide carrier, or by removing the bellows. In the latter case the cell is placed on a frame near the space occupied by the slide carrier. Such reactions as acids with bases with an indicator for color change, and metals and metallic oxides with dilute acid may be used.

Special Arrangements

The horizontal projection lantern has the disadvantage of being placed at some distance from the front of the room, often at an inconvenient position in the center of the room where fixed seats or tables may interfere, or where facilities are not available. Certain arrangements are possible which tend to overcome this disadvantage.

Double Reflection

Two mirrors are arranged to reverse the direction of the light. It is necessary to use mirrors of good quality in order to avoid a loss of intensity of illumination. However, if care is used, this arrangement is satisfactory. It has the advantages of placing the projector near the front of the room and of inverting the image to be observed, so that where a projection cell is used the normal position of the contents is observed on the screen.

Use of Short-focus Lens

The horizontal projector may be equipped with a short-focus lens. The usual lens provides for a projection distance of 20 ft to 35 ft. If this can be replaced with a short-focus lens (approximately 4 in.), the projector may be used near the front of the room in one of the following ways: The projector may be placed within the demonstration desk or table. It is ready for use by lifting a section of the table top and needs no adjustment. It is necessary to set the projector at a slight angle, but if care is used this angle does not interfere seriously with the operation of the bulb.

The section of the table top may be hinged. The prolector is mounted on the underside of this section. A translucent screen is mounted inside the table and pulled up so that it stands between the projector and the observers. If it is not possible to mount the projector within the desk, it may be placed on the top of the desk. An opaque screen is placed in the corner of the room.

The horizontal projector may be used for illumination only, serving simply as a system of condensers. The object to be projected is placed somewhat beyond the projection lens. Another lens is placed beyond the object to focus the image of the object on the screen. This arrangement is quite satisfactory, particularly where cells and other objects are too thick to be placed in the normal projection position.

It is not necessary to depend upon the horizontal projector as a source of illumination. Such illumination may be provided by the use of an arc lamp or strong projection bulb. A stage or space for objects is provided and the projection lens placed beyond this. Such an arrangement may be made permanent, thus freeing the horizontal projector for other uses.

Built-in Projector

An arrangement which is essentially an overhead projector is built into the demonstration desk. The stage of the projector is flush with the surface of the demonstration table. All that is necessary for projection is the placing of the mirror in position. This is stored conveniently within the desk. The stage is protected when not in use by a plate or may be made of a piece of plate glass in. in thickness to withstand the wear that it normally would receive as a portion of the surface of the demonstration table. It is essential that provision be made for cooling of the projection bulb.

Handmade Slides

While prepared slides may be secured from a number of different sources, it is often desirable to prepare slides by hand to meet local needs. Plain lantern slides (generally listed as "lanternslide cover glasses") are nominal in cost. While they may be cut to size (3 in. × 4 in.) from window or other glass, this is generally not satisfactory. If the glass is dirty it should be cleaned thoroughly. If soap and water and other common methods fail, the cleaning solution of potassium dichromate may be used. The slides should be rinsed thoroughly with water after the cleaning.

It is possible to write directly upon a clean glass slide with India ink or with the waterproof type. This may be satisfactory for a slide or two, especially if permanence is not an important factor. Usually it is difficult to obtain very consistent and permanent results without coating the slide. Particular care should be taken that the fingers do not touch the glass where the writing with ink is to be, because the perspiration makes it difficult for the ink to adhere to the glass. It is possible to obtain ink specially prepared for writing on glass slides.

Coating Glass Slides

Two simple and inexpensive methods have been found to be particularly usable for coating slides. These are solutions of gelatin in water, and of clear lacquer in lacquer thinner. A small amount of the solute in the solvent is satisfactory--about i part to 50 or even 100 parts in either case.

For coating with gelatin, a small amount of granulated gelatin is obtained. This may be purchased from a drugstore or laboratory supply house. Ordinary clear or plain dessert gelatin may also be used. The latter may be purchased from grocery stores. Approximately ¼ teaspoonful of gelatin is dissolved in a cupful of hot water--or 5 ml or 10 ml of gelatin to 500 ml of water. The solution is placed in a vessel large enough to cover a slide easily, and the slide, held by its edges, is dipped in the solution. The slide is supported at an angle on paper towels, newspaper, or blotting paper, to drain and dry. It may be written on in 5 min or 10 min.

For coating with lacquer, a solution of I part clear lacquer to 50 or 100 parts of lacquer thinner is used. These may be purchased from hardware or paint stores. This solution may be applied to the slide with a small brush, or the slide may be dipped. The slides should be placed at a slight angle with the table top, resting one edge of the glass slide on a meter stick or long fiat piece of wood, with the opposite edge resting on the table. Newspapers or other protection should be placed under the slides. The lacquer evaporates rapidly and the slide is ready for use almost immediately.

The dipped slides have the advantage that both sides are coated so that either side may be used. With slides which have been coated on only one side, as those coated with lacquer and a brush, it may be desirable to indicate in some way which is the coated side. This may be done by placing a dot of India ink in one corner of the slide.

One can write or draw very easily with India ink upon the coated slides. Care should be taken that the lines are rather fine; a crow-quill pen which may be purchased from stationery stores is satisfactory. The slide may be placed over squared or graph paper, so that straight and uniform lines are made. A margin of approximately ⅜ in. on each edge should be maintained, so that all writing and drawing show on the screen.

Ground-glass or Etched Slides

Ground-glass slides can be written or drawn upon with pencil or India ink using a crow-quill pen. Lantern-slide ink which works satisfactorily with ground-glass slides may be purchased.

Abrasives which are suitable and easily obtained for making ground glass slides are fine emery or Carborundum powder (No. 120 or finer). Valve-grinding compound (for automobile motor valves) may be used. The water-base type is more satisfactory than the oil-base type. Other commercial abrasives are available. A small amount of the abrasive should be placed on a smooth piece of iron or on a piece of glass plate somewhat larger than a glass slide. Water is added to moisten the abrasive. The glass slide is placed over this mixture and ground with a rotary motion. It should be frequently examined until it is ground over the entire surface, and care should be exercised that it is not ground too much. Continued grinding makes the slide less transparent with less satisfactory results.

The main advantage in using ground-glass slides is that they may be written upon with an ordinary pencil and colored with either special lantern-slide crayons or lantern-slide inks. The colored inks give a more brilliant color. The chief disadvantage of the ground-glass slide is that it reduces the illumination significantly.

Any type of glass slide may be easily treated so that it may be used again. Uncoated slides which have been written upon with India ink and ground- or etched-glass slides which have been drawn upon with crayons may be washed with warm water and soap. A pencil or ink eraser may be used on ground-glass slides which have been written upon with ordinary lead pencil. This may be followed by the warm water and soap. Etched-glass slides which have been written upon with the special lantern-slide inks may be cleaned with the special solvent which is furnished with the inks. Used glass slides coated with gelatin may be cleaned with warm water, while those coated with lacquer should be cleaned with the lacquer thinner. The plain glass slides then should be cleaned before they are recoated or used again.

Protecting Prepared Glass Slides

The prepared slide may be used in the projector as soon as dry. Continued use soon may scratch and dull the slide unless it is protected. An effective way of protecting prepared glass slides is the placing of another clean slide cover glass over the prepared slide and binding the two together with slide binding tape. A disadvantage of this method of protection is that it uses two cover glasses for each completed slide. The material on the slide is thoroughly protected, however.

Another satisfactory way to protect glass slides is by applying a coating of lacquer directly over the prepared slide. This works particularly well with the uncoated slides or with those coated with the lacquer solution. The lacquer should be more concentrated than that for the undercoating previously used. A satisfactory proportion is one part of the clear lacquer to one part of the thinner. The lacquer may be put on in the more concentrated form. Care should be taken in applying the lacquer that it flows onto the slide without having to work the brush back and forth over it, otherwise the ink will run. The ink should be dry, and there should be a good supply of solution on the brush.

If the slide is valued it may be desirable to use the extra cover glass and bind the two slides in the regular way. Where permanence is not important the lacquer coating is quite satisfactory. Boxes having partitions to separate glass slides from each other are available, or may be constructed. Some types have grooves into which slides may be placed. Such boxes are rather bulky, because of the unused space needed for separating the slides. Another practical limitation is in identifying each slide. The slides may be lifted or each may be numbered, with a key to those numbers provided.

A method which has proved satisfactory consists in placing the slides in envelopes and writing on the upper edge of the envelope a notation concerning the nature of the slide. The slides may be placed closely together without danger of scratching, and arranged with guide cards in a regular card file. The notations at the top of the envelopes can be read easily as the slides are separated. Slides of a unit may be placed behind a labeled guide card without regard to a number scheme and found easily when desired. A satisfactory envelope for this purpose is the manila pocket used by librarians for the usual book card.

Making Cellophane Slides

For the use of typewritten material, the cellophane slide is excellent and is easily and economically prepared. Cellophane slides, cut to size and inserted between a double carbon paper, may be purchased from various commercial sources. Ordinary wrapping cellophane may be utilized; it is cut to size (3¼ in. × 4 in.) with shears. It may be purchased in a variety of colors.

The cellophane is placed between a double carbon paper of good quality which is made twice the size of the slide and folded in the middle. The cellophane is placed inside with both carbon sides against it, so that the typing will occur on both faces of the cellophane. A 3/8-in. margin should be maintained all around the cellophane when typing. The ribbon should be set as for stencil cutting so the type does not strike through it, and the carbon paper should be used only once to ensure good impressions.

Cellophane may be used for lettering or drawing with India ink, though considerable care must be used as the sheets are very thin and a pen point easily pricks through and a blot occurs. Heavier cellophane than that for wrapping is commercially available. India ink may be used directly upon such slides without the use of a coating. Care should be exercised in drawing or writing upon cellophane that the fingers do not come in contact with the cellophane, as the perspiration makes it difficult for the ink to adhere. A mat cut from heavy cardboard or metal with an opening approximately 2½ in. × 8 in. is placed over the cellophane slide. The writing is done through the opening, while the slide is held firmly by pressing on the margins of the mat.

When the cellophane slide is completed it may be placed between hinged slides for projection. The cellophane slide should not remain in lantern with a strong light source (500-w bulb or more) for more than a few minutes. The cellophane slide may be bound between two cover slides with tape for permanent use or it may be filed in manila library pockets.

Filmstrips

The filmstrip is known by various names such as the strip film and slide film. The filmstrip consists of a series of pictures or other materials on a strip of 35-mm film. Each of the pictures or frames is projected in turn, thus producing results comparable with those secured by the use of the lantern slide. Filmstrip projection has the advantage of relatively low cost of both the projector and the filmstrip. The projector is simple to operate and readily portable. The filmstrips are relatively unbreakable, although they may be damaged. There is some disadvantage in their fixed sequence, although this is generally not serious.

There is an abundance of filmstrips of interest to the science teacher which cover many areas of science in detail. Each filmstrip may have from a few frames to one hundred or more. The content of the filmstrip may vary greatly in detail and in the maturity level for which it is adapted, so the science teacher should study the filmstrip carefully in advance and use those frames which serve his purpose best. The filmstrip may be used in its entirety or selected frames may be used.

Filmstrips may or may not be titled. Some of the filmstrips are made to be used only with a manual and, therefore, have no titles; others have titles but not manuals. The titles accompany the pictures or frames in some filmstrips; in others they are on separate frames. The teacher should be thoroughly familiar with the content of the titles, and manuals, and the nature of the pictures before he starts planning for their use.

The projection of filmstrips requires either a separate projector or an attachment for a projector for slides or opaque materials. There are commercially produced combination projectors for filmstrips and 2 in. × 2 in. slides.

The use of a filmstrip projector has additional advantage for the amateur photographer, who can make his own pictures with a camera using 35-mm film. The advantages rest not only in the fact that the teacher can make the exact pictures which he wishes but also in the ease of making the pictures. The teacher may, if he wishes, develop his own pictures, thus making the process even more direct and less expensive.

The production of such visual materials by the science teacher opens up an abundance of materials not otherwise available. Suggested uses are: the photographing of industrial operations (it may be necessary to secure permission), laboratory equipment, pictures, charts, and the like which are otherwise inaccessible for class use because of their size or location.

Opaque Projection

The purpose of opaque projection is to place on a screen an enlargement of opaque materials. A room to be used for opaque projection should be relatively darker than is necessary for such techniques as projection of film slides and lantern slides; however, the abundance of materials which may be secured is such that the effort needed to provide adequate darkening of the room is well spent. These materials may in clude: pictures, graphs, charts, maps, real objects, printed material to be read from books, magazines, and newspapers.

Current and back issues of magazines and newspapers have an abundance of material which the science teacher may wish to use in enlarged form for the class. Likewise, texts and reference books, laboratory manuals, trade journals, and photographs may be used.

Typed materials such as tests and laboratory directions may be projected, also. Hand-drawn charts, diagrams, and maps may be presented by the same method. Where it is feasible, it is well to type or draw the material to be used on Bristol board or other heavy-weight paper or cardboard. The images of real objects may be projected providing those parts which are to be observed lie in approximately the same focal plane. The image of various small materials may be projected readily on the blackboard. The image on the blackboard, which has served as a screen, is traced. The details needed are included and those not needed may be omitted as the situation requires.

Opaque projectors are available commercially and provide satisfactory results. If it is not possible to purchase a commercially made device, a satisfactory projector can be built in the laboratory or shop. The area to be projected is limited to dimensions of approximately 6 in. × 6 in. depending upon the projection lens used. It may be constructed from sheet iron of No. 22 or No. 24 gage. Plywood lined with sheet asbestos may be used instead of sheet iron.

The dimensions suggested are not critical except for the determination of the depth from the lens to the picture to be projected. This depth should be slightly greater than the focal length of the lens. Because of the variability of the distance from the projector to the screen, it is essential that the lens be made adjustable. This may be provided by the use of a rack and pinion which is used on the lantern-slide projector, or the lens may be mounted on the end of a cardboard mailing tube by the use of Scotch or adhesive tape. This mailing tube fits snugly inside another mailing tube and is held in place by the friction between the two. It may be adjusted easily for focus. A third device for adjusting the lens is a vertically mounted rod; the lens is adjusted by a clamp which is mounted on the rod. The mirror may be supported likewise on such a rod or other vertical support. It is essential that the mirror be held in a swivel clamp so that its position may be changed to raise or lower the picture on the screen.

The bulbs used should provide rather strong illumination on the projected material. Two 200-w bulbs will provide adequate illumination for most purposes. If a 500-w bulb is used, it should be placed with its base down and ventilation should be provided. A small electric fan which forces air through the apparatus serves quite well. If less intense illumination is satisfactory, the forced draft is not necessary. For such illumination, holes in the bottom and top portions of the illuminating chamber to provide for convection currents are adequate. However, it is necessary that shields be provided so that the direct illumination from the interior of the box does not escape into the darkened room.

Certain variations in the construction may be provided with legs and a door in the bottom held in place with a spring. The door should be about 1 ft on each edge. This serves to support any materials so that a fiat surface is exposed to the illumination. The spring should be strong enough to prevent the weight of the material from causing the door to remain partially open when the material is in place.

A variation of the opaque projector may be constructed in either vertical or horizontal form. This device has the advantage of eliminating the door and the adjustment that it requires, although adjustment of the lens is generally necessary after the projector has been put into place. If it is used in a horizontal position, the materials are held before the opening for projection. In the horizontal position the mirror may be removed for more intense illumination, but the reversal of the image by the mirror is lost, and this is not so satisfactory, particularly where printed materials are projected.

Chapter 7
CONCLUSION

Several institutional barriers affect collaboration in higher education. A major barrier is the way courses are designed. Course content is separated into discrete, subject specific areas belonging to a particular department (Blenkinsop & Bailey, 2002; Brownell, Yeager, Rennells, & Riley, 2000). A second barrier is that faculty members teach in one department only, usually where they hold tenure. Faculty report feeling penalized for their attempts to team-teach or pursue other collaborative opportunities (Sapon-Shevin, 2003).

Philosophical differences among faculty may provide yet another barrier to successful collaboration (Brownell, Yeager, Rennels, & Riley, 2000; Bondy, Ross, Sindelar, & Griffin, 2002). In their pilot test of a collaborative course, two professors from different disciplines found that they often disagreed on how to approach the course, the students, and the idea of collaborating in the classroom (Bowles, 2004). As with faculty in most departments, there were disagreements as to how much to collaborate, how much to integrate, and what approach would be best for all.

Further, there may exist an actual physical location barrier to collaboration among faculty in higher education. Sapon-Shevin (2003) explained that many colleges of education separate special education and regular education departments on different floors and even in different buildings. This is an issue at the University of Wisconsin-Eau Claire, where the general education and special education buildings are separated by a river. Faculty members and students must cross a bridge to get to the other side for one of their special education or general education classes, and these two academic buildings are the two farthest apart on campus.

In addition, the structure of the university reward system often impedes collaborative efforts among faculty. Shared and individual research receives more emphasis within the reward structure in colleges of education than collaborative research. Moreover, faculty involved in collaborative teaching and research may not have the support of their personnel committees when decisions for promotion and tenure are made. Therefore, there is a lack of reward structure for faculty members who work collaboratively (Brownell, Yeager, Rennells, & Riley, 2000; Fauske, 2002).

Communication is a critical element in the collaborative process. This is an element in the collaborative process that is often overlooked. In their study of an integrative course for science and language arts methods Blenkinsop and Bailey (2002) indicated that it was the way in which they communicated that nurtured and maintained the collaborative relationship. Furthermore, Fauske (2002) in her study of cross-campus collaboration explained that one condition for sustaining collaboration is to establish structures for facilitating communication. Sapon-Shevin (2003) described that lack of communication between special education and general education faculty perpetuated feelings of distrust and fear and further obscured the dialogue necessary for effective collaboration. For any collaborative effort between faculty to be successful, faculty must learn how to effectively communicate.

Collaborative efforts among faculty in higher education can be successful and may reap many rewards for faculty (Austin, & Baldwin, 2001; Gray, 2000; Blenkinsop & Bailey, 2002; Quinlan, 2004; Brownell, Yeager, Rennells, & Riley, 2000; Bowles, 2004; Martin, 2002) if faculty are attuned to the benefits. Gaining knowledge and new insights into teaching and student learning is one benefit of faculty collaboration. Martin (2002) examined teacher collaboration involving the restructuring of curriculum and how teachers involved in collaborative efforts gained valuable insights into their own teaching. The teachers involved in the case study reported that they shared professional knowledge and the development of significant new understandings and insights into their own teaching. Brownell, Yeager, Rennells, and Riley (2000) cite several studies demonstrating collaborative efforts, and fostering a sense of community and a shared vision among faculty. University faculty view themselves as more knowledgeable, gaining new insights into teaching and learning because of collaborative efforts.

Faculty collaboration creates connections between departments and divisions. Fauske (2002) found that among a department of teacher education, a department of English, and four secondary schools, faculty realized that they shared strong beliefs about preparation of teachers. In an example of a collaborative effort among several departments of a college of business, Quinlan (2004) found that the project leaders believed that their efforts created better understandings between departments where there had been distrust. Weaver and Landers (2001) found that both regular and special education faculty realized there was common ground between the two programs that was necessary for all their students in their review of a new teacher education program.

The faculty collaboration can provide a vehicle for building relationships among faculty (Koop, 2004). Blenkinsop and Bailey (2002) reported that they connected as colleagues and developed a solid relationship through their collaborative teaching. Quinlan (2004) described how discussions among faculty about students, learning, and teaching can "recreate collegiality" (p. 45). Through collaboration, faculty can share burdens and pressures and support one another when necessary (Brownell, Yeager, Rennells, and Riley, 2000). Collaboration among faculty in higher education can build bridges between departments and divisions, reveal new ways of teaching and learning, and foster positive relationships for all involved.

In 2003 the Elementary Education Department and the Secondary and Continuing Education Department were merged and a new department, the Department of Curriculum and Instruction, was formed. Therefore, three departments preparing teachers (Elementary, Secondary, Special) became two: Curriculum and Instruction and Special Education. Although collaboration between the Elementary and Secondary/Continuing Education Departments did occur prior to the merger, the need to collaborate programs and curriculum became politically, economically and organizationally critical in 2003.

Another important merger that provides important context for the course discussed in this article is the 2004 merger of the Colleges of Education and Nursing with the newly created School of Human Science and Services into a single College, the College of Professional Studies. Moving from the design of colleges to the concept of three schools created opportunities for sharing faculty positions, interdisciplinary search committees, curriculum revisions, and program responsiveness to student need. One such example is the capstone experience described herein.

Curriculum reform efforts during the 2001-2000 academic year provided the Academic Policies Committee with the opportunity to move and vote on several recommendations including the following recommendations that were critical to the evolution of the Capstone courses discussed here. * The School of Education directs the School of Education Policies Committee to coordinate the development of a common student teaching seminar. The School of Education should adopt a general methods component for teaching diverse learners within the Core Teacher Education Framework for all education students.

During the 2000-2004 academic year, three faculty members were assigned the task of preparing a proposal for the "new" capstone experience across all three Departments in the School of Education Faculty. In the fall of 2004, all School of Education faculty approved this proposal after the "new" capstone completed its first pilot year.

"Capstone" is now popularly applied in teacher preparation programs as an adjective describing the activities, experiences and/or courses that make up the final touches in a pre-service teacher's university program (Weaver & Landers, 2001). Faculty consensus around a common vision about the make up of the most important element in a teacher education program is essential. The university experience should consist of meaningful, relevant, rigorous experiences and products culminating in a student's decision about choosing a job (e.g. grade level, school district size, etc.) and the faculty's decision regarding the student's ability to succeed in the teaching profession.

Integrated capstone experiences and courses have responded to this concern and have gained in popularity as teacher education programs changed to meet the needs of an ever-increasing, non-traditional student population with a diverse array of educational needs. Modular coursework, distance education, self-contained course orientation, competency based programming, individualized sequencing of courses, and application of life experience through portfolio review have all become popular means for moving students towards successful and timely completion of degrees in education with certification to teach in the public schools. As Howey (2001) explained, "A series of loosely coupled courses in professional education culminating in an even more disconnected experience, commonly referred to as 'student teaching' obviously cannot be the standard for professional preparation" (p. 143). The faculty in the School of Education at University of Wisconsin-Eau Claire believed more synthesis of the student teaching experience was necessary.

Student teaching seminars were historically designed to support and enhance the field-based student teaching or internship experience. The professional semester was viewed as a forum for strengthening emerging skills and dispositions to which students had been exposed, and had practiced in less intensive practical field experiences. It was important to the entire faculty of the School of Education that the practical aspects of student teaching were complemented with time for reflection upon the practices and principles explored in methods courses and their emerging philosophies penned by students in foundations courses. Also, the professional semester was envisioned as a bridge for pre-service teachers as they made their final progression from a student-orientation to a professional-orientation. Therefore the seminar faculty cadre carefully planned for a balance of required experiences with opportunities for students to employ personal choice in how they accessed the supports necessary for demonstrating their ability to put knowledge into real-life practice.

The student teachers and interns receiving certification in special education and regular education now attend a similar "capstone" seminar experience. The current "capstone" seminar experience meets four full Fridays during the professional semester, requiring student teachers and interns to leave their field placements and meet on campus.

Instead of providing a completely required program of content, the three faculty representing the departments of Special Education, Curriculum and Instruction, and Foundations in Education have determined the essential content for all education students receiving certification through the university. This essential content takes up only part of the class meeting time. The remaining class time is devoted to a "conference style format", allowing students to choose what topics interest them in their professional development.

The Field Placement Coordinator, and coordinator of the new format seminar, in collaboration with the other two faculty members, solicited on-campus faculty and cooperating teachers to present topics of interest to student teachers and interns during the seminar time. On-campus faculty and cooperating teachers responded with presentations including topics such as: classroom management, parent involvement, child abuse, strategies for gifted and talented, national teacher certification, substitute teacher ideas, multiple intelligences, collaborative efforts between regular and special education teachers and more. Work sessions for seminar assignments were interspersed as well. The seminar faculty then set up a schedule enabling student teachers and interns to choose conference sessions they wished to attend. Furthermore, the seminar coordinator sent a program of conference sessions to the student teacher/intern and their cooperating teachers. This served as a reminder and allowed them to discuss what sessions would be most beneficial to the student teachers/interns.

To model a professional conference, the seminar faculty prepared individual folders containing programs and other necessary handouts for the day. The students arrived early Friday morning, picked up their folders, and registered for their sessions. Furthermore, coffee, juice, and donuts were served to allow students and faculty time to informally discuss their student teaching/interning experiences. All teacher education faculty were invited and encouraged to attend this early session to talk with the students and show support of this "capstone" experience.

This new format of the capstone experience allowed the individual needs of the Special Education, Curriculum and Instruction, and the Foundations of Education departments to be retained, and also allowed student teachers and interns to choose professional enrichment activities that would enhance their field placements. In addition, on-campus faculty and off-campus cooperating teachers became part of this capstone experience, creating a true partnership between the university and K-12 schools. Finally, student teachers and interns gained a better understanding of attending a professional conference, making choices for professional development, and meeting other professionals in the field.

Since the goal of this pilot project was to better meet the needs of student teachers and interns, the input from the students themselves was most important. In anticipation of the new format, the capstone faculty had the students involved in the former format evaluate the course and the proposed changes for the new course. The majority of student teachers/interns believed a conference style format would better meet their needs. Comments such as "more options would allow us to attend the session we consider important and beneficial" were common. In addition, students indicated that smaller group sizes would allow for more discussion and problem solving. Finally, one student commented that "empowerment feeds motivation." After reviewing the evaluation from that semester, the capstone faculty felt confident with piloting the collaborative, conference-style format.

Educational reform at the university level requires vision, creative and committed faculty, supportive administration and plain hard work. The institution must remain acutely aware of how the structure of departments, schools, and colleges inadvertently undermine attempts to expand vision into creative action. Administrative support that goes beyond memos and policy must be provided especially to new faculty charged with the pioneering work. Respect of the efforts of the past must be demonstrated so that essential content and existing practice is not sacrificed just for the sake of change. When all these components are in place, innovative and meaningful changes grounded in historical relevance and professional trust can occur and be sustained.

Students, cooperating teachers, and faculty evaluative comments and behavior speak towards the value of the process and the efficacy of structuring decisions that led to the new capstone seminar. There continues to be on-going willingness of faculty across departments and disciplines to share a session during the conference style seminars. Furthermore, during the last semester over forty cooperating teachers volunteered to present a session and attend other sessions during the seminar. The common threads determined in the capstone course analysis are being traced back into prerequisite courses, and faculty are talking across departments about how to coherently build these common threads, while simultaneously weaving a rich diversity of thought and practice.

These efforts have not gone unnoticed by our students who watch faculty closely for insights into collaborative teaching. Students have noticed a change in faculty. They see us talking with each other and walking with each other engaged in reflective conversation about teaching practice. Thus, these steps towards meaningful reform have allowed us as faculty to do for our students what Robert Coles (2000) calls teachers to do for their students when he reminds us, "character is ultimately who we are expressed in action, in how we live, in what we do, and so the students around us know: they absorb and take stock of what they observe, namely us - we adults living and doing things in a certain spirit, getting on with others in our various ways" (p. 199).

Collaboration is emergent (Friend & Bursick, 2001). This means that as professionals engage in the dimensions of collaborative partnerships, they get better at collaboration. This was evident in the process undertaken at the University of Wisconsin-Eau Claire. The faculty members feel that their willingness to participate has increased, comfort level with theoretical and methodological divergence has improved, and enthusiasm for the work involved has been rekindled. Participation in the process is motivating and exciting. Student enthusiasm is contagious. Faculty members recognize the benefits of working together.

REFERENCES

Abraham M. (2000). "Research on instructional strategies". Journal of College Science Teaching, 18, 185-187.

-----. (2004). "Doing research on college science instruction: Designing experiments". Journal of College Science Teaching, 24, 150-153.

American Association for the Advancement of Science (AAAS). (2000). Project 2061: Science for all Americans. Washington, DC: AAAS. (Reprint: Rutherford J. F., & Ahlgren A. (2003). Science for all Americans. New York: Oxford University Press.)

-----. (2003). The liberal art of science. Washington, DC: AAAS.

American Chemical Society. (2001, July 8). Education policies for national survival. Washington, DC: American Chemical Society.

Austin, A.E., & Baldwin, R.G. (2001). Faculty collaboration: Enhancing the quality of scholarship and teaching. Washington, DC: Association for the Study of Higher Education. (ERIC Document Reproduction Service No. ED 346 805)

Blenkinsop. S., & Bailey, P, (2002). An inquiry into collaboration and subject area integration in teacher education. San Francisco, CA: American Educational Research Association. (ERIC Document Reproduction Service No. ED 388 649)

Bodner G. M. (1999). "Constructivism: A theory of knowledge". Journal of Chemical Education, 63, 873-878.

Bondy, E., Ross, D.D., Sindelar, P.T., & Griffin, C. (2002). Elementary and special educators learning to work together: Team building processes. Teacher Education and Special Education 18(2), 91-101.

Bowles, P.D. (2004). The collaboration of two professors from two disparate disciplines: What it has taught us. San Diego, CA: Paper presented at the Annual Meeting of the National Reading Conference. (ERIC Document Reproduction Service No. ED 386 744)

Brownell, E.Y., Yeager, E., Rennells, M.S., & Riley, T. (2000). Teachers working together: What teacher educators and researchers should know. Teacher Education and Special Education 29(4), 340-359.

Coles, R. (2000). How to raise a moral child: The moral intelligence of children. New York: Random House.

Fauske, J.R. (2002). Five conditions for sustaining cross campus collaboration on teaching and learning. (ERIC Document Reproduction Service No. ED 376 116)

Friend, M., & Bursuck, W.D. (2001). Including students with special needs: A practical guide for classroom teachers. Boston: Allyn and Bacon.

Gray, B. (2000). Collaborating: Finding common ground for multiparty problems. San Francisco: Jossey-Bass.

Howey, K. (2001). Designing coherent and effective teacher education programs. In J Sikula (Ed.), Handbook of research on teacher education (pp. 143-169). New York: Macmillan.

Koop, A.J. (2004). Empowering teacher educators: A process of transition. Brisbane, Australia: Paper presented at the Annual Conference of the Australian Teacher Education Association. (ERIC Document Reproduction Service No. ED 375 106)

Martin, K.M. (2002). Teachers' collaborative curriculum deliberations. San Francisco, CA: Paper presented at the Annual Meeting of the American Educational Research Association. (ERIC Document Reproduction Service No. ED 388 646)

Quinlan, K.M. (2004). Promoting faculty learning about collaborative teaching. College Teaching 46(2), 43-47.

Sapon-Shevin, M. (2003). Teacher education which promotes the merger of regular and special education: Challenges and opportunities. In H.S. Schwartz (Ed.) Collaboration: Building common agendas (pp. 115-123). Washington DC: American Association of Colleges for Teacher Education.

Weaver, R., & Landers, M.F. (2001). Preparing preservice secondary teachers for the diversity presented by students with special needs. Chicago, IL: Paper presented at the Annual Meeting of the American Association of Colleges for Teacher Education. (ERIC Document Reproduction Service No. ED 394 930)

Finster D. C. (2000, 2001). "Developmental instruction. Parts 1 and 2". Journal of Chemical Education, 66, 659-661; 68, 752-756.

Herron J. D. (1997). "Piaget for chemists". Journal of Chemical Education, 52, 146 - 150.

Katz D. A. (2001). "Science demonstrations, experiments, and resources". Journal of Chemical Education, 68, 235 - 244.

Moore J. W., Crosby G. A., Smith S. G., & Lagowski J. J. (2000). "Chemistry plus technology plus teachers yields curricular change: The FIPSE lectures in chemistry." Journal of Chemical Education, 66, 3 - 19.

Perry W. G., Jr. (2000). Forms of intellectual and ethical development in the college years: A scheme. New York: Holt, Rinehart and Winston.

Project Kaleidoscope Report. (2001). What works: Building natural science communities. Washington, DC: Project Kaleidoscope.

Schwartz A. T., et al. (2004). Chemistry in context: Applying chemistry to society. Dubuque, IA: William C. Brown.

Shakhashiri B. Z. (2001, December). "Achieving scientific literacy". Chemical and Engineering News, pp. 47 - 48.

Spencer B. (2001). "What works in chemistry education". In Chemical manufacturers association catalyst award brochure (pp. 12-16). Washington, DC: Chemical Manufacturers Association.

Tobias S. (2003). They're not dumb, they're different: Stalking the second tier. Tucson, AZ: Research Corporation.

Calvin C. S., & Lasgowski J. J. (1998). "Effects of computer simulated or laboratory experiments and student aptitude on achievement and time in a college general chemistry laboratory course". Journal of Research in Science Teaching, 15, 455463.

Case C. L. (1998). "The influence of modified laboratory instruction on college student biology achievement". Journal of Research in Science Teaching, 17, 1-6.

Crow L. W. (2000). "The nature of critical thinking". Journal of College Science Teaching, 19, 114-116.

Curtis J. B. (1999). "Teaching college biology students the simple linear regression model using an interactive microcomputer graphics software package". Dissertation Abstracts International, 46(7), 1858A.

Davis W. E., & Black S. (2000). "Student opinion of the investigative laboratory format". Journal of College Science Teaching, 15, 187-189.

Fields S. C. (2000). "The effectiveness of traditional biological laboratory activities on the learning of formal concepts by non-formal operational students". Dissertation Abstracts International, 46, 114A.

Foley J. D. (1998). "Interfaces for advanced computing". Scientific American, 257, 127135.

Hall D. A., & McCurdy D. W. (2001). "A comparative study of a BSCS-style laboratory and a traditional laboratory approach on student achievement at two private liberal arts colleges". Journal of Research in Science Teaching, 27, 628-636.

Journet A. R.P., et al. (1998). "Studies on cognitive development in a non-majors investigative general biology laboratory". Abstracts of the Annual Meeting of the National Association for Research in Science Teaching.

Kern E. L., & Carpenter J. R. (2004). "Enhancement of student values, interests and attitudes in earth science through a field-oriented approach". Journal of Geological Education, 32(5), 675-683.

Kyle W. C., et al. (2000). "Assessing and analyzing the performance of students in college science laboratories". Journal of Research in Science Teaching, 16, 545-551.

Lawrenz F. (2000). "Aptitude treatment effects of laboratory grouping method for Students of differing reasoning ability". Journal of Research in Science Teaching, 16, 279-287.

Lawson A. E. (1998). "Relationships among level of intellectual development cognitive style, and grades in a college biology course". Science Education, 64, 95-102.

-----. (2000). "A review of research on formal reasoning and science teaching". Journal of Research in Science Teaching, 22, 659-717.

-----. (1999). "Student reasoning, concept acquisition and a theory of instruction". Journal of College Science Teaching, 17, 314-316.

Lawson A. E., Rissing S. W., & Faeth S. H. (2003). "An inquiry approach to nonmajors biology". Journal of College Science Teaching, 19, 340-346.

Lawson A. E., & Snitgen D. A. (1999). "Teaching formal reasoning in a college biology course for preservice teachers". Journal of Research in Science Teaching, 19, 233248.

Lawson A. E., & Wollman W. T. (1997). "Encouraging the transition from concrete to formal cognitive functioning: An experiment". Journal of Research in Science Teaching, 13, 413-430.

Leonard W. H. (1999). "An experimental study of a BSCS-style laboratory approach for university general biology". Journal of Research in Science Teaching, 20, 807814.

-----. (1998). "Interactive videodisc: Computer instruction of the future?" Collegiate Microcomputer, 5, 197-201.

-----. (1999). "What research says about biology laboratory instruction". The American Biology Teacher, 50, 303-306.

-----. (2000a). "An experimental test of an extended discretion laboratory approach for university general biology". Journal of Research in Science Teaching, 26, 79-91.

-----. (2000b). "A comparison of student reactions to instruction by interactive videodisc or conventional laboratory". Journal of Research in Science Teaching, 26, 95-104.

-----. (2000c). "Ten years of research on science laboratory instruction at the college level". Journal of College Science Teaching, 18, 303-306.

-----. (2003). "Computer-based technology for college science laboratory courses". Journal of College Science Teaching, 19, 210-211.

-----. (2001). "Uncookbooking your laboratory investigations". Journal of College Science Teaching, 21, 84-87.

-----. (2002). "A comparison of student performance by interactive videodisc versus conventional laboratory". Journal of Research in Science Teaching, 29, 93-102.

-----. (2002). "The trend toward research on the teaching/learning process: Asking the right questions". Journal of College Science Teaching, 23, 76-78.

Leonard W. H., Journet A., & Ecklund R. (1999). "Overcoming obstacles in teaching large-enrollment lab courses". The American Biology Teacher, 50, 23-28.

Miller D. G. (1999). "The integration of computer simulation into the community college biology laboratory". Dissertation Abstracts International, 47(6), 2106A.

Morgan R. M., et a. (1998). "A microcomputer exercise on muscle physiology". Journal of College Science Teaching, 17(10), 23-27.

National Research Council. (2003). Fulfilling the promise: Biology education in the nation's schools. Washington, DC: National Academy Press.

Nicklin R. C. (2000). "The computer as lab partner". Journal of College Science Teaching, 15(1), 31-35.

Nisbett R. E., Fong G. T., Lehman D. R., & Cheng P. W. (1998). "Teaching reasoning". Science, 238, 625-631.

Rhodes S. B. (1999). "A microcomputer kymograph". Journal of College Science Teaching, 15, 523-527.

Smith M. (in press). "Doing research on college science instruction: Planning data analysis". Journal of College Science Teaching, 24.

Spickler T. R. (2001). "An experiment on the efficacy of intuition development in improving higher levels of learning and reasoning in physical science". Dissertation Abstracts International, 1, 143A.

Stevens S. M. (2000). "Interactive computer/videodisc lessons and their effect on students' understanding of science". Abstracts of the Annual Meeting of the National Association for Research in Science Teaching.

Tofte W. L. (1999). "The comparative effectiveness of learning center and traditional approaches for a college introductory geology laboratory course". Dissertation Abstracts International, 43, 358A.

Vernier D. L. (1998). "How to build a better mousetrap". Portland, OR: Vernier Software.

Walkosz M., & Yeany R. H. (2001). "Effects of lab instruction emphasizing process skills on achievement of college students having different cognitive development levels". Abstracts of the Annual Meeting of the National Association for Research in Science Teaching.

Waugh M. L. (1998). "The influence of interactive videodisc simulations on student achievement in an introductory chemistry course". Abstracts of the Annual Meetings of the National Association for Research in Science Teaching.

Wright E. L. (1999). "Effect of intensive instruction in cue attendance on problem solving skills of preservice science methods students". Abstracts of the Annual Meeting of the National Association for Research in Science Teaching.

www.ingramcontent.com/pod-product-compliance
Lightning Source LLC
Chambersburg PA
CBHW030814180526
45163CB00003B/1281